DESIGNING
HEALTHY COMMUNITIES

DESIGNING
HEALTHY COMMUNITIES

RICHARD J. JACKSON

WITH STACY SINCLAIR

FOREWORD BY ANTHONY ITON

APHA PRESS

JOSSEY-BASS
A Wiley Imprint
www.josseybass.com

Published by Jossey-Bass
A Wiley Imprint
989 Market Street, San Francisco, CA 94103-1741—www.josseybass.com

Readers should be aware that Internet Web sites offered as citations and/or sources for further information may have changed or disappeared between the time this was written and when it is read.

Limit of Liability/Disclaimer of Warranty: While the publisher and author have used their best efforts in preparing this book, they make no representations or warranties with respect to the accuracy or completeness of the contents of this book and specifically disclaim any implied warranties of merchantability or fitness for a particular purpose. No warranty may be created or extended by sales representatives or written sales materials. The advice and strategies contained herein may not be suitable for your situation. You should consult with a professional where appropriate. Neither the publisher nor author shall be liable for any loss of profit or any other commercial damages, including but not limited to special, incidental, consequential, or other damages.

Jossey-Bass books and products are available through most bookstores. To contact Jossey-Bass directly call our Customer Care Department within the U.S. at 800-956-7739, outside the U.S. at 317-572-3986, or fax 317-572-4002.

Jossey-Bass also publishes its books in a variety of electronic formats. Some content that appears in print may not be available in electronic books.

Library of Congress Cataloging-in-Publication Data

Jackson, Richard J., 1945–
    Designing healthy communities/Richard J. Jackson with Stacy Sinclair.—1st ed.
      p. cm.
    Includes bibliographical references and index.
    ISBN 978-1-118-03366-1 (cloth); ISBN 978-1-118-12981-4 (ebk.); ISBN 978-1-118-12982-1 (ebk.); ISBN 978-1-118-12983-8 (ebk.)
      1. Sustainable urban development—United States.   2. City planning—Environmental aspects—United States.   3. Distributive justice—United States. I. Sinclair, Stacy. II. Title.
    HC110.E5J33   2012
    307.1'20973—dc23   2011025367

Printed in the United States of America

FIRST EDITION
HB Printing   10 9 8 7 6 5 4 3 2 1

CONTENTS

PART III. BE THE CHANGE YOU WANT TO SEE IN THE WORLD

FOREWORD

In a startling article in the *New England Journal of Medicine* in 2005, researchers predicted that life expectancy in the United States would decline early in this century, primarily due to the obesity epidemic. In 2007, researchers attributed a regional stagnation and decline in life expectancy in the Southeastern United States to obesity and related chronic disease risk factors. And now, just last week, for the first time in twenty-five years, the Centers for Disease Control and Prevention reported an actual decline in overall U.S. life expectancy. Although this decline may or may not represent the beginning of a new downward trend, one thing is for sure . . . Dr. Dick Jackson warned us.

Thirty years ago many of us thought nothing of frying up a half pound of bacon for the family breakfast. Today most of us wouldn't even think of doing that, given what we now know about risk factors for cardiovascular disease. However, most Americans live in cities and suburbs designed decades ago, before we knew the potential health consequences of these designs. Our car-focused community design has facilitated a burgeoning obesity epidemic. In their seminal book, *Urban Sprawl and Public Health*, Dick Jackson and his coauthors remind us that the "modern America of obesity, inactivity, depression,

and loss of community has not 'happened' to us. We legislated, subsidized, and planned it this way." Dick Jackson has become a public health evangelist, traveling the countryside, imploring us to look at the profound correlations between obesity and urban design and how our "vehicle miles traveled" have escalated exponentially just in order to navigate our pedestrian unfriendly, *obesogenic* (obesity-producing) neighborhoods. He highlights how our children are particularly vulnerable to these new environments. Like a true public health detective, Dick Jackson has been putting the puzzle together for us. We ignore his message at our peril.

Dick Jackson is unusual for a scientist. He is accessible. While his research is rigorously methodical, analytical, and detailed, his presentation is inspiring, humorous, philosophical, and often poetic. He leaves us hopeful and motivated to rise to the challenge of reimagining and redesigning our communities. He is also courageous, and at many points in his career Dick has chosen the difficult route of speaking truth to power, despite clear personal and professional consequences. For that he has earned enormous respect in the rank-and-file public health community and has received accolades and awards culminating in the Presidential Distinguished Service Award, awarded to individuals by the president of the United States for achievements "so outstanding" that the individual "is deserving of greater public recognition than that which can be accorded by the head of the department or agency in which he is employed."

At The California Endowment we seek to promote fundamental improvements in the health status of all Californians. We recognize that the twenty-first-century public health challenges of obesity, chronic disease, and profound health inequity are not simple medical challenges but implicate community design, public policy, and the very political process itself. Given this understanding,

we have embarked on a ten-year, fourteen-community initiative called Building Healthy Communities (BHC). This name is in part a tribute to the insights of public health visionaries like Dick Jackson, who have shown us that we have to literally rebuild and redesign our communities in order to achieve the kinds of environments that will allow us to live in harmony with our human health needs.

Anthony Iton, MD, MPH, JD
Senior Vice President, Building Healthy Communities
The California Endowment

PREFACE

THE PERFECT STORM

The United States is confronting a "perfect storm," one where three powerful threats are converging to create near-catastrophic conditions. The first threat is social: an aging population and a hurricane of chronic diseases. The second threat is environmental, with challenges ranging from the microscopic to the global, including resource depletion and global heating. And the third is economic, particularly the struggles of middle-class and working people in a stagnant and staggered economic system.

The health and social storm is formidable. Sixty-eight percent of Americans over age twenty are overweight or obese, and obesity in U.S. children and adolescents has tripled in just over a generation. One in three American children is overweight. These changes raise the risks of heart disease, stroke, and many other illnesses, and especially, an epidemic of diabetes. Becoming severely obese (body mass index greater than 35) increases the risk of diabetes 40-fold for a man and nearly 100-fold for a woman. Developing diabetes before the age of forty shortens life expectancy by nearly fifteen years, and diminishes the quality of life by twenty years. If current trends are not reversed, this could be the first generation of American children to have shorter life spans than their parents.

Not only is the obesity epidemic prematurely aging our population but we also seem to be enjoying our lives less. In the last ten years, antidepressants have become the second most prescribed medication in the United States, and the percentage of the population, including children, receiving them has doubled. And we, especially our children, are becoming increasingly unfit. Nearly three-quarters of our high school students are unable to pass simple fitness exams. Regular physical activity in natural settings is beneficial for children, yet more and more children confront environments generally hostile to walking, bicycling, and independent play. Many young teens live in isolated housing developments and, being unable to drive, have no access to shops, community centers, and public transportation—their *community*. This isolation puts them (and also adults who cannot drive) at increased risk of boredom and depression.

Of all the interventions that could counter these epidemics, none works as well as increased physical activity, particularly when carried out in contact with nature. Yet if we continue to design and build America in ways that are hostile to walkers and bicyclists—creating an America without parks and without safe routes for walking or biking to destinations such as schools and workplaces—we are unlikely to see better physical and mental health.

The environmental storm is equally formidable. In 1950, the greenhouse gas carbon dioxide was at a level of 305 parts per million; by 2011, this number had increased to 391 parts per million. This change has been accompanied by increasing acidity in the oceans, glacial melting, sea level rise, and global heating. About one-seventh of the carbon dioxide retention in our atmosphere is due to loss of trees and vegetation; the remainder is due to the burning of fossil fuels. Doctors know that a patient who retains too much carbon dioxide is in trouble. All signs tell us it is also bad for our planet globally. At the same time, resource extraction has become more difficult, expensive, and destructive. For example, a century ago a small team of men could drill an oil well. Today oil wells require a huge infrastructure and investment. And in that same span of time the global population has grown from 2 billion to 7 billion.

The economic storm is the third tempest making up our perfect storm. Americans' real income doubled between 1950 and 1980, but since that time, real income has "flat-lined" for all but the top 20 percent of the population. After-tax income for the top 20 percent has improved and for the top 1 percent it has doubled. The income gap grows ever wider. Millions of the jobs Americans once relied on are no longer available, and trillions of U.S. dollars are now being held by other nations. Although tax and other strategies seek to end this dolorous situation, little change can be expected when close to $200 billion a year is being used to buy foreign oil. At the same time, Americans spend proportionally more of their income on transportation than any other population in the world. How and where we build our dwellings drastically affects what we spend and how we live. We remain servants of the automobile.

BUILDING HEALTHY COMMUNITIES

I have come to realize that those of us who work in environmental health have focused too much on "parts per million" and "parts per billion" and on remote places, and have neglected the origin of environmental health—namely, that people's health is profoundly shaped by their immediate environments, the places and things we as communities build, and where we house people. How and what we now build is being determined by financial and social policies, outmoded housing codes, and car-loving design principles that have long undervalued the needs of the poor, the young, and those who do not or cannot drive, and that, as is now becoming apparent, are also failing to produce an environment that meets the needs of everyone else as well.

We as a species are very adaptable to the environment in which we find ourselves. In the United States, we have designed and constructed a car-centric environment and adapted our lives to it. In making this adaptation we have often overlooked the dangerous health, social, environmental, and economic consequences that come with it. But as human beings we are also able to shape our environment. We need to start thinking about communities that work for all of us, young and old; communities that support those who walk, bike, or drive; communities that don't demand that we own a car or burn large amounts of fossil fuels; communities that create good, local, and meaningful jobs, for example, in artisanal production, high-quality construction, urban farming, solar power adaptation, teaching, and service.

Our urban design, especially the buildings and roads we make and maintain, can give people access to the places that help them fill their life needs, including food, shelter, work, and health care—or take it away. This human-made environment, or *built environment*, is at the core of environmental health. This idea was obvious to everyone in the nineteenth century, a time when nearly everyone knew someone with tuberculosis and nearly every family had lost a child to a diarrheal illness.

Public health has traditionally associated the built environment with systems that address such issues as sanitation, water and air quality, pest control, lead paint poisoning in children, workplace safety, fire codes, toxic sites, and access for persons with disabilities. These issues are important, but they do not tell the whole story. We now realize that *how* we design the built environment

may hold tremendous potential for addressing—and it is to be hoped, *preventing*—many of the nation's current public health concerns. And, as an urgently needed by-product, better design may reduce health care costs, allowing us to divert those savings to redressing the equally challenging imbalances in the quality of education in our country.

Our environment is everything around us. If an embryo makes it to birth, the environment influences that person's health even more than his or her genetic makeup. That environment is physical, social, nutritional, economic, and behavioral. The built environment, the physical one that we humans make, comprises our houses and workplaces, streets and water systems, parks and play areas. Even on the water we are in human-made craft. Most of us spend 99 percent of our time in the built environment and only rarely can we experience wilderness. How we design, and what we then build, create, and destroy, has profound impacts on our health. This is obvious when we think about drinking water or sewer systems or strong, stable buildings and safe roadways, but the built environment is shaping our health in equally profound though less apparent ways. It determines where we can live, work, visit, walk, run, shop, worship, and play.

Health is defined as a state of complete physical and mental well-being. We cannot have physical well-being in dark, damp buildings or on dangerous streets. Most of us cannot have mental well-being without human and nature contact and supportive environments. I often think that a healthy community is not unlike a computer: it must have highly functional software *and* hardware, the operating system as well as the computer chips. A highly functional community must have its *software*—systems for education, justice, an economy, and human support—as well as its *hardware*—homes, workplaces, roads, and more. No community can function long if it lacks either a social or a built environment. The best communities offer both. In the words of Charleston mayor Joseph Riley, "what you need to do is make sure that the place in which you live, and your people live, is as nice as the places they would dream about visiting."

For a community to be healthy, it must have both functional social systems and functional built environments. In fact, the built environment becomes social policy in concrete, literally. A city that provides good transit in its poorer areas is enacting a social policy (a good one in my opinion). A contrasting social policy is made concrete when a city demolishes low-income housing to build a sports stadium with forty acres of parking that is to be used ten times a year, mostly by wealthier suburbanites. A city that fails to build *complete streets*—streets that accommodate all users, including walkers and bicyclists—is enacting a policy that fails to serve the third of the population that does not drive. Cities that create quality urban centers that nurture culture, diversity, exercise, and farmers' markets are exerting a social policy as well, a healthy one. What we build is not enough in itself to create a healthy environment, but it is an essential and critical element of that environment.

ABOUT THIS BOOK

This book is intended for those of us who are concerned about our communities and the world we are giving our children. As I explain more fully later, it is a companion to the public television series *Designing Healthy Communities*, but it can also be used alone. It is not a textbook for a course on the built environment and health but is intended instead to help community members with understanding their current built environment and recognizing a range of possibilities for changing it. Specialists in public health, architecture, and urban planning increasingly recognize the links between health and built environment, and the public and its leaders are now beginning to catch on as well, but they need to effectively push for the healthy changes we all need.

Part One of this book is devoted to the software of a healthy built environment. The chapters in this part describe the characteristics of healthy communities and look at why caring, love, and *caritas* are important elements in these communities. Our caring—for ourselves and our community and for future generations—must shape what we build. We need to put not only social policy into concrete; we also need to put our caring into concrete. Our current nonsustainable way of building and of depleting resources and our profligate use of energy are a form of generational child abuse. It is remarkable to me that people who would not dream of taking a loaf of bread away from a hungry child seem unaware of the generational abuse inherent in paving over fine farmland, leveling forests, and destroying the planet's thin atmospheric cover. So in the first part of this book, we will think about the linking of caring to health in our communities.

Part Two examines communities that are working to transform themselves into healthful physical environments. In Belmar, a development near Denver, Colorado, a dead shopping mall has been retrofitted to create a lively downtown with housing, retail, and recreation, and that area is linked via transit to Denver itself. In Prairie Crossing, Illinois, a subdivision has been created with a view toward sustaining maximum green space, local organic agriculture, and a strong social network for the people living there. An old city that has revitalized itself, Charleston, South Carolina, was fortunate enough to escape the post–World War II destruction other cities experienced from "urban renewal" centered on automobile driving, and has redeveloped a lively, charming, diverse, and economically viable downtown. I reflect on the importance of political leadership in bringing about this creation. In discussing Boulder, Colorado, I talk about the vision of this city that values physical activity and has committed itself to *active transportation*, particularly walking and biking. In Elgin, Illinois, a drive toward sustainability is revitalizing this former one-industry town that came on hard times; this effort is coming both from city leaders and from the base, namely high school students led by a charismatic teacher named Deb Perryman. In Oakland, California, a port city that faces the world as a point of international trade, residents have

suffered degraded air quality and disease increases and have not shared in the benefits of the port. And the last chapter in Part Two looks at Detroit (Motor City), the home of automobile-dependent America, which is confronting twenty-first-century realities of loss of industry, depopulation, and poverty, and asks what we can learn from urban homesteaders and local agriculture.

Part Three of this book investigates ways in which the average citizen can have a voice and can help to take charge of the future of his or her community and the world we all are giving to our children.

In preparing this book and the public television series, several colleagues and I conducted many interviews, mostly in the communities discussed in Part Two. Unless otherwise indicated, comments quoted here made by individuals who live in these communities and a few others are taken from these interviews, carried out by me and by Harry Wiland, Dale Bell, and Stacy Sinclair, in 2009 and 2010. These interviews provide an essential element of both the book and series.

Throughout, I provide photographs and figures to illustrate and supplement the discussion. In the Portfolio, in the center of the book, you will find a number of color photographs from the locations my team and I visited during the filming of the public television series. The plates show both scenes of desolation delineating the pathology of the built environment gone wrong and best practice examples of what is possible with proper planning and execution. At the end of the book, you will find the notes and an index.

THE TELEVISION SERIES

As I have mentioned, this book is a companion to the upcoming special public television series *Designing Healthy Communities*. Produced and directed by the Media Policy Center in Santa Monica, California, the series describes, in depth, how the design of the built environment affects our health, with an additional emphasis on inequities related to that design and on the need for social and environmental justice. In this series I explore how the built environment has contributed to the fact that *two-thirds* of Americans are overweight; seventy million are obese, including many of our most vulnerable children; and many individuals suffer from an array of other chronic but *preventable* diseases, including type 2 diabetes, high blood pressure, asthma, heart disease, and depression. These diseases cost billions of dollars annually as we try to treat them. The series looks *upstream* at the root causes of our malaise and highlights best practices that can be put into action and are based on the efforts of real people with compelling solutions.[1]

The television series offers insights on our challenges and examines how our focus on building our country to meet the needs of cars and not of people, and especially not of our children, has undermined our health, wealth, and communities. At the same time, this project seeks to identify hope and options in the face of these challenges. Just as we must be mindful of the world we are giving our children, so too should we be alert to the gifts our children are offering to us. Over and over again in making this series and in my teaching, I have been struck by the resilience, optimism, and energy of young people. The revitalization of our communities and our nation can come from our children, if we are wise enough to listen.

Many people and organizations now recognize how the existing built environment is harming public health and needs to be redesigned or retrofitted to reflect new information: for example, that toxic, contaminated air is a factor in asthma; that having insufficient sidewalks, bike paths, and parks is a factor in obesity; that a lack of access to fresh food and farmers' markets is a factor in diabetes; and so forth. I am convinced that the struggles and triumphs of concerned community activists, politicians, socially responsible businesses, and ordinary citizens presented in this book will engage, encourage, and inspire you to action, wherever you live, work, study, or play.

It is possible to weather a perfect storm, but not in poorly built structures, whether social or physical. A good solution solves multiple problems. A good solution to our social, health, environmental, and economic challenges is to design and build places that allow us to meet our life needs well, and to be with the people we love as much as we can, and to spend only as much time in cars as we really want to. We need to focus on this now, because even with the best will in the world, it won't happen overnight. Moreover, it won't happen at all unless we decide we want to give our children a better and more health-giving world.

ACKNOWLEDGMENTS

This book, along with the accompanying public television series, was instigated by the many people who have told me that its message—that the environments we build can greatly help, or profoundly harm, our health and happiness—needed to be more widely heard.

I need to first acknowledge the creativity, social awareness, support, and provocation of Harry Wiland and Dale Bell, who have been powerful tugs pulling and pushing this work along; the patience and diligence of the brilliant film editor, producer, and writer Beverly Baroff; and the research, diligence, camera work, and artistry of Jonathan Bell, Charla Barker, Teresa Chang, Scott Izen, Aaron Kemp, Adil Khanna, Alan Mabry, Troy Mathews, Karen Ng, Ruben Rajkowski, David Rosenstein, Sari Thayer, Brianna Tyson, and all the energetic staff of the Media Policy Center. I also most profoundly thank Stacy Sinclair for working with me on this book. She was superb in taking a mass of facts, insights, places, people, and stories, and organizing them into a logical and provocative structure. Her diligence and discipline, humor, and knowledge, especially in the area of education, educated and assisted me enormously. I now also see why she skippers a large sailboat in heavy seas so well.

My intellectual mates in the built environment and health voyage have been led by public health physicians Howard Frumkin, now of the University of Washington,

and Andrew Dannenberg, both formerly of the Centers for Disease Control and Prevention (CDC) and both remarkable leaders—brilliant, disciplined, and patient. I am grateful for the work of so many in the planning field, especially Lawrence Frank at the University of British Columbia, as well as my health and planning colleagues at the University of California, Los Angeles, and the University of California, Berkeley.

The reviewers and editors who reviewed my manuscript and provided me with valuable suggestions and constructive criticism were Parris Glendening, Leslie Meehan, Preston Schiller, and R. K. Stewart. I thank them for both their time and their feedback.

This work was first launched thanks to a seed grant from the board of directors of the American Institute of Architects. Essential to the provisioning of this voyage have been the Kresge Foundation, the California Endowment, the Kellogg Foundation, the Gifford Foundation, the Marisla Foundation, Kaiser Permanente, and the Robert Wood Johnson Foundation.

My colleagues at the CDC, the California Department of Public Health, UC Berkeley, and UCLA have been a huge help and an inspiration, and I thank them. I am especially grateful to my wonderful students. There are too many to name, but I must especially acknowledge Rachel Cushing, Heather Kuiper, Lisa Martin, Marlon Maus, Kyra Naumoff, Mathew Palmer, Tamanna Rahman, June Tester, and Andrew Tsiu.

My wife, Joan, and sons, Devin, Galen, and Brendan and his bride, Cheryl, are the anchors and tell-tales in my life, as are my six brothers and sisters and their families, and of course our patiently strong, red-haired mother, who gave me the gift of life, of insight, and of gratitude.

Los Angeles, California   Richard J. Jackson
July 2011

THE AUTHOR

Richard J. Jackson is professor and chair of the Department of Environmental Health Sciences at the School of Public Health at the University of California, Los Angeles. He has also served in many leadership medical positions with the California Department of Public Health, including the highest, state health officer. For nine years he was director of the Centers for Disease Control and Prevention's National Center for Environmental Health, in Atlanta.

His work in California led to the establishment of the California Birth Defects Monitoring Program and state and national laws that reduced risks, especially to farmworkers and children, from pesticides. At the Centers for Disease Control and Prevention (CDC) he established the national asthma epidemiology and control program, the environmental health tracking program, and built environment and health activities, and advanced the Childhood Lead Poisoning Prevention Program. He instituted the current federal effort to biomonitor chemical levels in the U.S. population. Jackson was the U.S. lead for government efforts around health and the environment in Russia under the Gore-Chernomyrdin efforts, including radiation threats. In the late 1990s, he was the CDC leader in establishing the U.S. National Pharmaceutical Stockpile to prepare for terrorism and other disasters—which was activated on September 11, 2001. He has received numerous awards, including the Breast Cancer Fund's Hero Award and the UC Berkeley School of Public Health's Distinguished Teacher and Mentor of the Year award. He has received lifetime achievement awards from New Partners for Smart Growth, Making Cities Livable, and the Public Health Law Association.

Jackson is a coauthor of *Urban Sprawl and Public Health* (with Howard Frumkin and Lawrence Frank, 2004) and is the host of an upcoming public television special on public spaces and public health, and coeditor of the follow-up textbook. He has testified before the U.S. Congress and the California legislature. He has served on many health and environment boards, numerous Institute of Medicine efforts, and the board of directors of the American Institute of Architects. He received his MD degree from the University of California, San Francisco, and did his pediatric training there as well, and his master of public health (MPH) degree, in epidemiology, at the University of California, Berkeley. He is board certified in pediatrics and in preventive medicine.

He is married to Joan Guilford, and they have three sons—the oldest recently became a physician epidemiologist at the CDC, the second is in public service work, and the third is in the arts, especially film, in Northern California.

To Robert Jackson, whose short life took me to public health;
to Dorothy, whose humanity took me to medicine;
to Joan, who gives me love and insight; and
to Brendan, Devin, and Galen, who inspire me

PROLOGUE

WHY I CARE ABOUT THE BUILT ENVIRONMENT

A few years ago I was on a morning radio talk show and the host challenged me. "You work for the government? You must be lazy or stupid or corrupt," he said. I responded, "No, I am none of those. I am a physician, a pediatrician, and I picked the career of public health to make a difference, to embrace life's challenges, not to control people's lives, but to assure conditions where people can be healthy."

As a public health officer, I use every tool I have learned in my training—medicine, pediatrics, epidemiology, statistics, toxicology, and psychiatry—but the most important tool I use is communication, in order to share information that is technically competent but also compassionate and honest. How we use our words, our lexicon, is important. Doctors call the tracking of disease *surveillance*, though that word has a very different meaning to the FBI. *Development* means child fulfillment to pediatricians, but to the State Department it means nation building. So in this book I am communicating ideas, and I hope the words I use are tools of health.

This book is intended to communicate some of the public health challenges that arise from our built environment, to help others see what I see, so we can all play a part in designing healthier communities for our children and grandchildren.

The *built environment* is everything we have made in order to live our lives. It is our homes, places of business, public spaces, and parks and recreational areas—or the lack thereof. It extends to electric transmission lines, waste disposal facilities, and transit pathways that keep us moving, communicating, and functioning. Every building and designed space we see was at one time a sketch in someone's imagination, and at some point a decision was made to build solely a functional structure or one that functions while it inspires.

The Golden Gate Bridge, in my opinion the most beautiful large bridge in the United States, was built at about the same time as the Empire State Building in New York City. The Grand Coulee Dam on the Columbia River was open by the beginning of World War II, irrigating the Northwest and bringing water and electrical power to war efforts. The theme here is capturing our culture in concrete. These iconic structures were created during an economic downturn, when labor and materials, including concrete, were less expensive and people needed work. The government put people to work creating not only functional structures but also magnificent gifts to our grandchildren and future generations.

The United States and other civilizations must work not just for the economy but also for people in communities that are stressed and in need of support. If we are going to make changes, we ought to be creating spaces that work for our health, the economy, and the planet—places that are *of the heart*. When we look around our communities, what we see are the choices people made in the past manifested as the communities in which we live now. We can choose to keep the parts that work for us and to change what does not.

We have paved over 60,000 square miles of the United States, an area the size of Georgia. We have gotten rid of trees and we have created impervious surfaces, with the result that rainwater and melting snow run off the streets and into concrete sluices rather than percolating into the soil to nourish and cool the planet. Then we buy bottled water because we are made to think that we do not like our local drinking water. We have more cars in the United States than we do licensed drivers. No wonder the parking lots and streets feel so congested—and we build this landscape at a cost.

I believe that *how* and *where* we build affects people's health. To build communities that are good for people will require new partnerships and new thinking. Children have value intrinsic to the gift that they are. Today's children deserve a planet at least as healthy, beautiful, and biodiverse as the one we received from our parents and grandparents.

In nature, everything is fascinating. Ecosystems are remarkably biodiverse yet stable and produce little waste. In an ecosystem, something is always consuming something else. There is a balance of old and young, big and small, with immense functional variety. A forest can sit on a patch of land for 80,000 years and the land will be richer for it, unlike what Western civilization has done to the land. Biological systems can digest sugars and proteins—all kinds of natural products. It was not until we began bonding large amounts of organic molecules to elements like chorine, fluorine, and bromine that molecules that could not be readily degraded in natural systems became pervasive in the biosphere, that is, in all the living things around us. These molecules, which have health effects, concentrate in the bodies of animals and humans, with the human infant at the top of a *bioconcentrating* food chain. Through our habits and practices, we humans are creating the next great mass extinction of species, comparable to the earth being hit by an asteroid. Nature does not tolerate dysfunctional systems for very long. Thanks to the astonishing molecule DNA, biological systems change and do so much faster than we once thought; indeed, evolutionists have been stunned to realize how quickly change happens. We are learning that our current exposures affect our own DNA and, in turn, these changes will affect not only us but also the generations who follow. As we look at our own communities, we must ask whether we are leaving the next generation a gift or a staggering burden.

GROWING UP

Children learn about the world around them through experience, and the primary providers of that experience are their parents and the other adults in the community. When I was little and lived in Portland, Maine, my father was an air traffic controller. When he took me for a walk he would say, "Stay close to me. There are bears in these woods." I remember picking blueberries and bringing back a bucketful. Maybe that is where I began to make the connection between love and belonging with the natural world and the connection between food and joy.

Figure P.1   Robert "Bobby" Jackson and his P-51 Mustang in Iwo Jima, September 1945.

Source: Photograph from the Jackson Family.

My father flew P-40 and P-51 fighter planes in the South Pacific during World War II (Figure P.1). He crashed at least once and survived a massacre while at Iwo Jima. He married the love of his life at twenty-two, my mother. After the war they had three sons in three years. I was the oldest son, named after my father's brother, a landing ship tank (LST) skipper stationed in the South Pacific.

On August 19, 1949, when I was three, polio stopped my father's breathing after two days of illness. After the funeral, when I still had no idea what death was, my grandmother told me, "You're the man of the family now. You have to take care of your mother and your brothers." That message has been a gift and burden.

Growing up I was a good reader, and I studied year-books, magazines, and pictures, especially pictures of my father. These images had a profound impression on me, and I stored the insights, memorable comments, and Irish family jokes of the people around me. I still remember when I was six, after my grandmother died, my mother saying, "the veil over the future is held by the angel of mercy." She tried to help me worry less about the future so I could focus on fulfilling the present.

My widowed mother remarried when I was eight years old, and then had four more children. The nine of us lived in a three-bedroom, one-bathroom house in Nutley, New Jersey. My stepdad struggled with poor health and was intermittently out of work. There just never seemed to be enough food in the house, especially with five boys, and we would go and get USDA surplus milk powder, macaroni, and other foods to help stretch our family's food budget. Dinner was a verbal free-for-all. There were many meals we barely could finish because we were laughing so much at the stories our youngest sister, Kathleen, would tell.

I joined the Jesuit seminary Saint Andrew-on-Hudson, in Poughkeepsie, New York, at age eighteen. I planned to be a priest. When I entered the seminary I weighed 132 pounds, and I must have gained a dozen pounds in the first two weeks I was there. I was delighted to eat whatever was put in front of me, and I did not mind that we ate in silence for many of the meals. My young Jesuit friends stunned me. No subject would come up, not a year in history, a thought leader, or a political event, on which someone at our table did not have a remarkably deep knowledge and the ability to provide commentary.

For two years I prayed five hours a day, lived mostly in silence, spoke in Latin, and learned the Jesuit life and the insights of Greek, Roman, Catholic, and modern philosophers. I also learned about Henry David Thoreau and read his book *Walden* about five times in two years. He spoke right to my core as he observed the beauty of the natural world. The Jesuits used to say *age quod agis*—"do what you are doing"—but I became restless. The seminary was intellectually stimulating, but it certainly did not meet all my needs as a young man. I left the seminary, completed a range of college science courses compressed into two years, and ended up in medical school in San Francisco.

PRACTICING MEDICINE

I loved the Bay Area, Yosemite, and Point Reyes. The city, the mountains, hiking, music, the bay, and the wonderful quality of life in San Francisco transfixed me. I married a beautiful woman, and I became a pediatrician. I graduated medical school in 1973 and completed my internship (I used to call it my internment as the hours were so long) and residency at the University of California, San Francisco, and San Francisco General Hospital.

During my residency I soon realized that to doctors a patient is defined by data—the statistics and lab results. To me the data are only a part of the picture. When I was working at the San Francisco VA Hospital, I got to know and like an old veteran who talked a lot about World War I. I had grown up with stories about WWII and had not heard these earlier stories. Sixty years after the war, he would still have nightmares about being gassed, and his stories moved me. One day I had to present his case to the attending internist, and of course no one wanted to know his fascinating stories. He was a *case* to be solved, not a *person* to know. He died after I had known him for a few weeks, and being an ex-seminarian, I wanted to go to his funeral. I was told, "doctors don't do that," with the clear message that I was not going to amount to much of one.

When I was in my third month of internship in pediatrics, a delightful fifteen-year-old girl came in for chemotherapy for leukemia. She was a very lively and bright girl, who had taught herself American Sign Language so she and her friends could gossip at school soundlessly during class. I was glad to see her on morning and evening rounds. She seemed to be in stable health when I left on Saturday, but when I returned on Monday, I learned she had died. I was instructed to go to the morgue in the basement to observe the autopsy. Though I had known all along that she would eventually die of leukemia, I learned that she had died more directly from the chemotherapy we gave her.

I remember going home feeling very low and realizing that there are no safety nets for emotions in medicine. There is an ugly saying in medicine that "you learn the most from the patients you kill." No one makes mistakes on purpose, but as horrible as it sounds, doctors do learn from each case. I was doing what I was told and administering the medications according to protocol, but this experience taught me about myself, and my limits. I stuck with my residency but surprised myself at the end when I told my favorite and most supportive cardiology professor, "Dr. S., I am not going to do an infectious disease fellowship. I'm going to do public health." He sat back in his chair and sighed. "Where did we fail?" he asked. I loved medicine, but I was also very intuitive, social minded, and political. I realized I could do more good in public health than I could seeing one patient after another. I also think that my medical school professors did not fail.

LEAVING TRADITIONAL MEDICINE

There is a long tradition of service to our country in my family. My great-great-grandfather served in the Union Army throughout the Civil War. His father was an officer in the War of 1812 and his uncles had served during the American Revolutionary War. My father and uncle served in World War II. My brother Jim became a Marine and served sixteen months near the Ho Chi Minh Trail in Vietnam during some of the worst fighting. At the same time, my brother Bill served in the Air Force. I was deferred from service until 1975, when I served in the U.S. Public Health Service, as does my son Brendan now.

I went to the Center for Disease Control (CDC), as the agency was then known, and was assigned as an epidemiologist to the State of New York. It was an amazing experience. I was working at least one epidemic per week, with terrific staff, and I never knew what was going to happen next. The most remarkable case was a cluster of hundreds of school children with severe bellyaches and fevers leading to sixteen appendectomies. We discovered that the cause of illness was the production procedure used by the dairy supplying the school. The dairy was adding chocolate syrup to already pasteurized milk under unsanitary conditions. When we stopped the chocolate milk supply to the school, the epidemic stopped. The bacteria that caused the illness, *Yersinia enterocolitica*, grew beautifully in sugary, alkaline, cold solutions, such as refrigerated chocolate milk.

The CDC assigned me to work for three months for the World Health Organization. In 1976, I was sent to Bihar State in India to work on smallpox eradication. I was in my twenties and I had 400 to 500 people working for me doing *containments* for suspected smallpox outbreaks. I am very proud to have helped in eradicating that terrible disease. There is a power to working in public health—one person or a small team can make a world of difference.

After working for the CDC and finishing my service, I returned to the Bay Area where I went to work as a pediatrician and also studied more epidemiology, especially as it involved chemicals. I wanted to help farmworkers and was fortunate to take on a position at the California Public Health Department as an epidemiologist looking at the health impacts of pesticides. It took me close to ten years to learn the hundreds of chemicals, often used in combination and with an ocean of brand names, the varieties of toxicity, and the dozens of health effects involved in pesticides. Some of the chemicals had toxicity close to that of nerve gas. In the process I learned not just about toxicology but also about worker health, social justice, air and water pollution, regulatory schemes, food contamination, and a *whole lot* of complicated, bare-knuckle politics (remember, I was in California). I learned that many of the pesticides had been legislatively grandfathered, meaning that if they were in use when the original pesticide law had been passed, they needed little additional testing for continued use. As a result, the data gaps were stunning. I knew this, and some of the people in the state agriculture department knew this too because we had access to the trade secret files that the public did not. I found that there was no tracking of birth defects in the state and no cancer registry in the larger agricultural areas.

In one case, I was investigating a baby boy with *tetramelia* (the absence of the arms and legs) born to a farmworker. We tried to learn what the mother had been exposed to, but it was impossible. The mother could scarcely remember the farms where she had worked, especially during the vulnerable first two months of pregnancy. Farmers had to keep records of use only for the most toxic chemicals. For weed and fungus killers, the chemicals more likely to cause birth defects, there was no record keeping required. These chemicals tended to be not immediately toxic. As a result, there were few toxicology data, negligible record keeping of pesticide use, and no tracking of birth defects.

With the help of lawyer friends and the American Academy of Pediatrics, I helped to draft bills to address the tracking issues, but they were quickly "killed" by the political process. The agricultural community saw adding these regulations as an unnecessary expense. I felt helpless until the medfly came to the rescue.

In 1980, the medfly threatened a large portion of California agriculture, one of the largest economies in the United States. The remedy was to spray the pesticide malathion over areas where infestations had been detected.

Unfortunately, most of these areas were urban and sub-urban. The public and the governor were deeply opposed to spraying over heavily populated areas. Because of the enormous outcry from the public, I was told to convene a health advisory committee on the issue. In the course of briefing the committee and while discussing the toxic-ity of malathion, issues stemming from data gaps and the lack of disease tracking emerged. These discussions became a powerful impetus for fixing the data gap problems and this led to support, even among members of the California legislature's agriculture committees, for addressing health concerns and establishing disease registries.

There is also a larger message in this story. Sometimes elected officials, the public, and doctors and scientists do not pay enough attention to a problem until it reaches the point of crisis, yet that moment of impending crisis is extremely important and must be used as a *teachable moment*, to create conditions in which people and com-munities can be healthier.

ESTABLISHING THE CONNECTION BETWEEN ENVIRONMENT AND HEALTH

In 1994, I moved back to the CDC to become direc-tor of the National Center for Environmental Health (NCEH). By this time the CDC had undergone a name change, becoming the Centers for Disease Control *and*

Prevention, because we have learned that we must do much more than control disease—we have to move upstream to prevent it. I believe that a critical way to do this is by changing the environment.

In April 1997, President Clinton issued an Executive Order requiring protection of children's environmental health and safety. Because of my work on pesticides and children's environmental health issues, I was named to lead this effort for the U.S. Department of Health and Human Services. Ten federal agencies were challenged to work together in ways that were unprecedented outside of wartime. Although most of the group wanted to focus on their own specialized diseases or toxic substances, I felt that the effort should focus on dealing with the larger issues affecting children's health.

After much discussion, and not a little argument, we focused on preventing developmental disorders (includ-ing lead poisoning and birth defects), preventing injuries (a major health problem in children and a major cause of children's death), reducing asthma morbidity (the most common chronic disease in children), and establishing a children's research effort, a longitudinal cohort study now known as the National Children's Study. There had never before been a long-term, longitudinal study on the impact of environmental factors on children from concep-tion through adulthood. This study remains in progress and promises important and groundbreaking information for the fields of medicine and public health.

In many ways, what came out of this effort was unprecedented. By focusing on, for example, the prevention of childhood asthma, we could move beyond academic battles over which air pollutant was most important and turf battles over which agency was making the greatest contribution. I wanted to see these conflicts fall into the background as each of the departments unified around a goal to protect and promote children's health. This experience was for me both an inspiration and a goal for what I think we must do with the redesign of our communities in the United States. By making our goal the improvement of children's health and well-being, and setting aside the turf battles over zoning, planning, and air and water pollution, we can focus on giving our children healthy places, and we can offer solutions that work for future generations—and perhaps for some of us alive today too.

Frederick Law Olmsted is one of my heroes in this regard. He is considered the father of landscape architecture, and he was the designer of Central Park in New York (Figure P.2), the Emerald Necklace in Boston, and many other parks that make urban living enjoyable and even delightful in the United States. Olmsted did not recognize that social health and personal health were separate. He never would have accepted the idea that mental health and physical health were divorced. He also served as the head of the U.S. Sanitary Commission created by Abraham Lincoln during the Civil War. The work of the commission, building and maintaining the hospitals for the wounded Union soldiers, employed new methods that vastly improved survival rates. Olmsted's policies made sure that hospitals—including the food and water, wounds

and bandages—were clean. He and many other heroes of the time met the fundamental goal of public health, "to assure the conditions where people can be healthy" or, in this case, can regain their health. What the commission did was simple, elegant, and it worked. My friend Larry Cohen, the founder and director of the Prevention Institute in Oakland, California, says, "a good solution solves multiple problems." Sanitation is one example of this. A safe, green neighborhood park, what Olmsted called the "lungs of the city," is another. It clears the air; captures rainwater; reduces pollution; raises property values; and improves physical, mental, and social well-being.

Figure P.2   Olmsted realized his idea of a public park as a green space accessible to all citizens in his design of Central Park.
Source: Flickr; photograph by David Shankbone. Used with permission.

Early on in my work at the NCEH, I realized that the CDC's activities were poorly integrated with each other, in a way sometimes described as *stove-piped* or *siloed*. As a result, those of us who practice public health, in order to confront and reduce diseases, have not been dealing with some of the upstream environmental health threats that cause lung disease, injuries, depression, and many other health problems. We need to identify good, smart solutions that cross multiple domains. Communities must put in place business and taxation interventions and new building codes and laws. And this nation needs to shift its awareness of what constitutes a good quality of life. Advocating for and working toward these changes may not sound very interesting, but for too long we have had doctors talking only to doctors, and urban planners, architects, and builders talking only to themselves. The point is that all of us, including those in public health, have got to get out of the silos we have created, and we have got to connect—actually talk to each other *before* and *while* we do our work—because there is no other way we can create the environment we want. Public health in particular must be interdisciplinary, for no professional category owns public health or is legitimately excused from it. It is also critical that public health have strong links to the medical world.

When we think about community design, it is like ecological thinking. There is a question of spatial scale. Some issues are small scale, like the way a porch is situated on the front of a house; some issues are intermediate scale, like the way a neighborhood looks (are there good sidewalks, stores within walking distance of homes, or a nearby park?); and then there are issues on the large scale, like an entire metropolitan area (is there a mass transit system, and is there a concern with striking a balance between constructing roads and making trolleys, trains, and buses available?). It may be relatively new to most people, including those in public health, to think in design and geographical terms, yet as all of us who are concerned about our communities come to think like geographers, we can think about different spatial scales and we can focus in on what we can do at each spatial scale to create the safest, healthiest places possible.

CONSIDERING THE FUTURE

Nowadays I am much more focused on chronic diseases, and today our society is facing an avalanche of them, especially diseases related to obesity. Doctors frequently measure health and examine patterns. They use the data and their experience to determine what will happen if patterns of health and sickness remain unchanged. When children are seen in a clinic, their height and weight are tracked by age to determine if their growth is on a healthy track. When I teach, I tell a story that I made up but that every health care provider knows is far too common. A ten-year-old child comes in for a physical exam. For his age, he is in the 50th percentile (right in the middle) for height and in the *95th percentile* for weight.

He is much too heavy, and his blood pressure is too high, which is probably related to his weight. His blood sugar is too high, but he is not diabetic yet. His cholesterol is too high, and he seems kind of depressed. When the physician talks to him, she learns he is lonely, he is not very good at sports, and life isn't going very well. She wants to help him but knows the success rate with these cases is not high (Figure P.3).

What is she supposed to do? She gives advice and refers him to a nutritionist. She says, "Mom, no soft drinks in the house. Get the television out of the bedroom. Let's make some changes in his life. Let's get him into an exercise program." She also says, "Make sure he doesn't bypass physical education," knowing that a lot of parents permit this because children who are not in great shape don't enjoy it. Two months later the child comes

Figure P.3   Patient and doctor talk about changing habits.
Source: Photograph from the Media Policy Center.

back and says, "I lost one pound but there's no place to walk. They pick me up and take me to school on the bus. The only real exercise I get is at school, and I can't control what they're feeding us in the cafeteria."

Two months after that, the child is medicated for elevated cholesterol, glucose, and blood pressure, and perhaps for depression. How did this problem start, and is it really his fault that he's not getting well on his own? This is, at its core, a toxic event, but is it the result of personal choice or due to greed, lack of knowledge, and pollutants in the environment? I assert that the built environment is rigged against this boy. It is also rigged against the doctor. The cumulative impact of this injustice, if it is not remedied, will crush the United States.

Let's go back to this same patient. What if he, with the support of his family, decides he is going to bicycle or walk the one mile to school four days a week? That would be about a thirty-minute trip on foot and would consume about 125 calories each way, or 1,000 calories a week. After a year he has burned 40,000 more calories than he did in previous years and is doing better in school because he has gotten some exercise. He has lost eleven pounds of body fat in one year. He goes back to the doctor at the end of two years. He has grown four inches (because kids grow about two inches a year). He is now down to the 65th percentile for weight. His blood pressure, cholesterol, and sugar are now normal. His energy level is good. His walking has also saved about 1,200 miles on the family car (assuming the parent drives a round trip morning and evening) and about

$600 in car costs. He has made new friends and perhaps has a girlfriend. If we multiply this scenario times fifteen million children, what we get is a sea change. This is not an absurd thing to ask. It is how the United States was in 1960.

The more I looked at chronic disease and epidemic issues, the more I realized that I needed to look upstream. Policymakers do not often design communities and buildings. In 2003, I gave a talk to about 500 professionals from the American Institute of Architects (AIA). I told them how important physical activity is and how they needed to design buildings with beautiful and attractive stairways so when we enter the buildings we can walk up and down the stairs. I asked them "to stop putting the stairs in some dark and dirty corner, key-controlled, too narrow and unlit, that are kind of scary. People want and need to use the stairs and it's good for health. It will help us lose weight and strengthen our bones. I want people to take stairs at least once a day." Afterward, I was worried that some important architect would corner me and berate me for not knowing anything about architecture, telling me that I should keep quiet. But instead, what the AIA president said to me was, "Doc, we love what you're saying. We love vertical features. We can have much more fun designing interesting spaces with stairways. Get the fire marshals off our backs, but what you are asking us to do is what we really want to do."

The AIA put me on its board of directors for a couple of years. I had a great time. (I suspect that architects want to be doctors and doctors want to be architects.) Kaiser Permanente is also taking on this challenge by running an ad that encourages people to walk up stairs (Figure P.4). It is a beautiful picture, though I would imagine it is somewhat hard to walk up stairs with no handrails while wearing high heels.

There are great ideas everywhere. For example, Gary Cohen of Boston's Health Care Without Harm has been working with green hospital-building projects. Energy-efficient hospitals are being built in Europe and around the world. When case studies of these projects are compiled, they will start cross-pollinating good ideas for healing and environmentally responsible medicine. As the scientific evidence accumulates, we are gathering very good indications that people travel, live, and play differently when the design of their built environment reflects connectivity, density, and mixed use.

For a while I was chairman of the American Academy of Pediatrics Committee on Environmental Health, and recently I helped to produce a policy statement from the academy about designing communities to promote physical activity in children. It took time to get this statement through four committees and one hundred chapter chairpersons. At first there was some pushback; members asked why pediatricians would issue a statement offering guidance on urban planning. The statement, now approved and made public, asserts that children need to grow up in environments where they can have increased physical activity and autonomy. I frequently have colleagues come up to me now to tell me that this statement just makes good clinical sense.

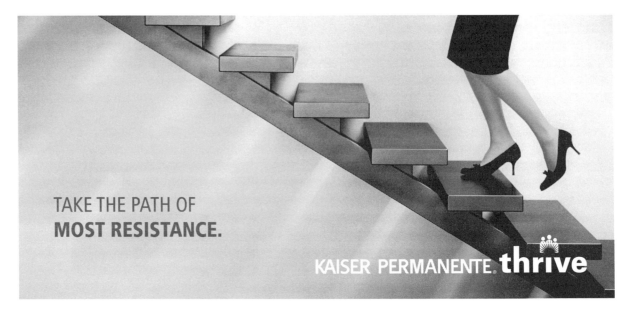

Figure P.4 A poster to encourage taking the stairs.
Source: Courtesy of Kaiser Permanente.

When I was a young pediatrician, my colleagues and I never saw a child with "adult onset" diabetes. It is now called type 2 diabetes, and my endocrinology colleagues at the University of California, Los Angeles, tell me that currently about half the children seen in the diabetes clinic have type 2 diabetes. Young people in their late twenties and early thirties with diabetes are now developing diseases that we think of as associated with old age. At the rate that the epidemic of obesity and diabetes is progressing in the United States and worldwide, it is clear that if we do nothing, about a third of our children will become diabetic at some time in their lives, with a reduction in their average life span of fifteen years and a reduction in their quality of life of about twenty years.[1] Someone who has gangrene from diabetes and is going to dialysis three times a week does not have a good quality of life.

We are also on a trajectory of declining life expectancy for the first time in our country's history. When I was a young pediatrician, people thought that having 7 percent of all the money in the United States going to medical care was a staggering amount. That amount is now at 17 percent.[2]

We ought to be the healthiest people in the world given our economic investment, but we are actually about dead-middle in terms of health statistics worldwide. I think this outcome has to do with the difference between *medical* care and *health* care. We are not creating health and we do not create well-being simply by medicating people at the ends of their lives. We are adding millions of jobs in the health care industry, but we cannot build an economy based on medical care. Eventually we have to grow things and produce products that add to the authentic wealth of the nation.

The public health community is starting to look at these issues. One sign of this is that at the American Public Health Association Annual Meeting in 2002 there were no abstracts of papers on land-use issues. In 2005 there were 55 abstracts, in 2008 there were 82, and in 2009 there were 130.

If we are going to build health, then we need healthy communities. Many places are working on this, with varying degrees of success. Most impressive so far are efforts in Boulder, Colorado; Charleston, South Carolina; and Portland, Oregon; and similar efforts are also increasing in big cities like San Francisco and New York.

The following chapters offer three distinct conversations. Chapters One, Two, and Three discuss the lenses through which we view the built environment and the ways that environment affects our health. Chapters Four through Ten describe seven places around North America that are making notable efforts to improve community health. And Chapters Eleven, Twelve, and Thirteen detail ways in which the reader can use this information to improve his or her local community. Through your own efforts you can join in—and even improve upon—the kinds of positive health impacts being created by the organizations and individuals whose stories you are about to read.

DESIGNING
HEALTHY COMMUNITIES

PART ONE

HEALTH AND THE BUILT ENVIRONMENT: AN INTRODUCTION

From the beginning, we have built environments to support or hinder our way of life. What we build, what those buildings are made of, and how they are designed and situated in space all reflect some of our most basic ideas about humanity in general and our specific relationships to each other. To understand what it means to design healthy communities, we must understand that each of us has a unique lens through which we see the world and interpret our surroundings.

When I travel I take great pleasure in looking at the buildings in the communities I visit. I marvel at the strength, beauty, and meaning that specific structures and combinations of buildings and open space in communities hold. I choose to visit communities because of how they *feel*. There are places that feel important, romantic, or beautiful because of what they are as well as because of the historical events that occurred there.

Part One of *Designing Healthy Communities* explores each word in the title of the book. The first chapter deconstructs what I mean by the built environment and the impetus for and the importance of the choices we make when we construct spaces. Our connections to *place* are strong, and they come from a feeling of love (or its absence). When we make design choices, these decisions are based on values and beliefs. When we look at existing environments, we can view them through this lens to understand what was important to those who last envisioned the space. Acting on our beliefs, we can choose to preserve important spaces or redesign them into a new vision for our future. Both preservation and redesign send messages about the impermanence of buildings and the changing nature of humanity.

Chapter Two examines what I mean when I talk about health. I began as a pediatrician, but I learned through a variety of experiences that our health is more than the absence of disease. Health is the result of a complex series of choices. Our surroundings and the resources available to us influence these choices. When we look at the leading causes of death now versus those a hundred years ago, we can understand the importance of personal choice and the outcomes that result.

Chapter Three reflects on the idea of community and the complexity of being a part of many communities simultaneously. There are some communities we are born into and others we strive to access. Communities are the people we interact with, yet groups of people live, work, and recreate in spaces, and the spaces take on a personality reflecting group interactions and challenges.

What if we looked at our lives through the lenses of love, our health, and the communities we engage in? What could we learn about ourselves and the choices we make for our built environment? I invite you to take this journey with me—to discover how each filter we use to look at our world can color our perceptions of it and how this can lead to new ideas for making our built environment a more desirable place to live in.

Chapter 1

What Does Love, or *Caritas*, Have to Do with the Built Environment?

Love—for all the use this word gets, it is still a word that gets lost in our roiling culture. *Ubi caritas et amor Deus ibi est*. This is a classic hymn sung in Latin, written nearly two millennia ago. It translates to, "where charity and love prevail, there is God." The absence of charity and love is the absence of God—a lonely place in any religion. Without God, one must struggle with existential loneliness. The Greeks had specific words for different kinds of love: *éros* for sexual love, desire, and the creative urge; *philia* for loyalty between friends and family, or enjoyment of an activity; and *storge* for natural affection, as between parents and children. There are many different kinds of love, but the word that matters in this book is *agape* in Greek, or in Latin *caritas*, the origin of our word *charity*—and this kind of love resides in our will, not in our emotions.

For goodness sake, what is this discussion doing in the story of built environment and health? Because, I think, the built environment is the embodiment of what we love, our imagination, and our will. It is what we value and reflects what and whom we care about. When it comes to love, talk is cheap—every adult has learned this lesson. It is what someone does, not what he or she says. And the built environment is very much what we do—in concrete.

When we meet a genuinely giving person, and we comment about the nature of her relationship with others, she invariably answers, "Oh no, I get more than I give." So too, I think, do we from really great built environments—Chartres Cathedral with a beauty that has astonished for a thousand years; the Golden Gate Bridge with both its profound utility and its elegant, simple beauty that is, amazingly, an improvement on a land form of breathtaking vistas; or a simple seacoast cabin in Maine without a single knickknack or unneeded ornamentation. I think that when we design and construct environments that give joy and comfort, safety and beauty, we get places that reflect the best of ourselves in which to live our lives.

I grew up in Newark, New Jersey, near the Passaic River, one of the largest toxic waste sites in the United States. About once a week my mom would bring my brothers and me across the Jersey meadows to see her sister and her sister's family. The ride through the swamps amazed me. We shut the car windows as we saw piles of yellow powder burning or smelled the stench of rotting garbage in the landfill.

When I went to study in the seminary, I found that even though the Greek, German, and modern philosophers had remarkable insights, the philosopher who really made sense was a peculiar and tubercular New Englander named Thoreau. The seminary was a place of high intellectual stimulation, but the sensual and physical worlds were marginalized, which was very difficult for me as a young, curious man of eighteen. I discovered that Thoreau spoke right to my core as he observed the beauty of the natural world—Walden Pond, the land, weather, and time. Oddly, once I better understood the natural world, I could better understand the textures and fragrances of the human world.

I went to medical school because I wanted a life of service and believed I would get meaning in my life out of that service. We were saturated in medical school with negatives, things to worry about—bad news and diseases. I learned a lot of pathology, and a lot about medicines. We were taught virtually nothing about health. I guess my teachers thought we would get that elsewhere. I became an epidemiologist and chased diseases. Epidemics of infectious diseases, pesticide poisonings, cancer, and birth defects became my life's

work. I soon realized I had to begin to learn less about disease and a whole lot more about the embedded health in the world around me if I was to make an impact.

I worry that in a larger way we are training our children the way we train doctors—lots of pathology, too little health—and too little time for independence and time with nature. I fear our children lose a lot when trash-culture, machines, and electronics dominate their lives. Realizing that well-built environments enhance and protect the natural environment, not erode it, and then providing those environments is an important way of revitalizing our children and their childhoods.

WE LOVE OUR FAMILIES AND OUR COUNTRY, BUT DO WE REALLY LOVE OURSELVES?

When looking at health data, we have to wonder, do we really care about *us*? Do we care enough to eat well and make healthy choices? Food should be a good sensory experience, from selecting ingredients to preparing a dish to consuming it. Eating involves a blending of color, texture, temperature, taste, smell, and even sound. It can be a social experience, and food choices also reflect our cultural identity. So why don't we eat better? We keep blaming individuals who get fat and do not eat sensibly, but the truth is we have a powerful national apparatus that is designed to make us fat and unhealthy (Figure 1.1).

Figure 1.1 Fast-food choices.

Source: Wikimedia Commons; photograph posted by Christian Cable. Used with permission.

We eat poorly because the food that is bad for us is cheap and abundant, tastes pretty good, is easy to get, and is full of salt, sugar, and fat. Taking the time to purchase, prepare, and enjoy healthy food requires a mind-set of self-caring. Although we might exert the effort for others, we often will not do it for ourselves, preferring to "pick up something quick" or "eat on the run" rather than to prepare, present, and patiently enjoy a meal alone. Doctors and nurses are often the worst offenders.

The food that is good for us tends to be perishable, can be comparatively costly, is often unfamiliar to most of us, and requires more attention and development than sugar, salt, and fat. Although we are subjected to limited educational efforts telling us to eat healthy food, the whole marketing system is directing us, is pushing us, toward unhealthy food. Obesity is on the rise, and children are building unhealthy eating habits from a very early age. Childhood obesity has increased to 12.4 percent for every age group (Figure 1.2). Whereas children ages two to five and twelve to nineteen are seeing some decline in obesity recently, the six- to eleven-year-olds are surpassing the rest of the young population, with nearly 20 percent of these children being obese.[1]

Looking at 2008 data from the Centers for Disease Control and Prevention (CDC), reveals that African Americans were 51 percent and Hispanics were 21 percent more likely to be obese than the white population in the same geographical area was. Obesity is more prevalent in the South and Midwest of our country. The highest rates of obesity (greater than 30 percent of the population) and diabetes (greater than 10 percent of the population) are found in the Southeast, Appalachia, and some tribal lands in the West and the Northern Plains.[2] (See the obesity map in the color Portfolio, Plate 1, for more information.)

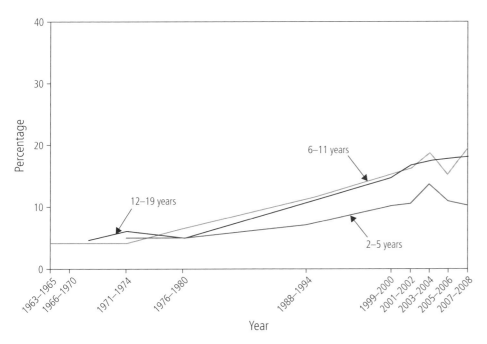

Figure 1.2 Obesity trends among children and adolescents: United States, 1963–2008.

Note: Obesity is defined as a body mass index (BMI) greater than or equal to the sex- and age-specific 95th percentile on the 2000 CDC Growth Charts.

Source: C. Ogden and M. Carroll, *Prevalence of Obesity Among Children and Adolescents: United States, Trends 1963–1965 Through 2007–2008* (Atlanta: Centers for Disease Control and Prevention, National Center for Health Statistics, 2010), http://www.cdc.gov/nchs/data/hestat/obesity_child_07_08/obesity_child_07_08.pdf.

When we live apart from what we know we need or rightly deserve, we are restless, anxious, and sometimes alienated, and this can lead to stress disorders and depression. Separating people from real needs and fair shares robs not just the individual of his own full potential, it also steals from his loved ones and the rest of society.

FOR LOVE OF FAMILY

Hippocrates, the father of medicine, reflecting on his profession, observed that "life is short, art is long." In medicine, we are constantly confronted by both the fragility of life and the inexorability of death. I revere architecture and building as art "in concrete," and this art is "long" and it far outlasts us if it is good. We are merely the custodians of the spaces where we are, including our homes. Buildings should have longer lives than we do. They are a legacy. Even so, many of today's buildings are created with a projected fifteen- to thirty-year life span—designed to be treated as consumer items rather than *real* estate. This is bad in terms of our environment, economy, and energy use. Our homes belong to the next generations too.

Think of Leonardo da Vinci's famous five-hundred-year-old drawing *Vitruvian Man* (Figure 1.3), named for Vitruvius who lived 2,000 years ago and is considered the father of architecture. Vitruvius was both a builder and an architect, and his precept was that a building must have *firmitas, utilitas, venustas*—solidity, usefulness, and beauty. Da Vinci's famous drawing illustrating the symmetry of the human body—strong, responsive, beautiful—captures a link between buildings and our bodies but also suggests the important idea that buildings must be adapted to human scale. The famous Danish architect and urban planner Jan Gehl suggests that many of the problems of modernist architecture and building design arise because it is not designed for people who inhabit these spaces in day-to-day life but rather is designed to reflect the ideas and ideals of the designer. Gehl suggests that too often modern design looks wonderful from an airplane or as we look down on a model in the studio—sighting it from above—but it just does not work at the human level, the places where we spend our lives. He asserts that places that look humdrum from above, even for example, Greenwich Village in New York City, are engrossingly vibrant at the human scale (Figure 1.4).

Perhaps this loss of human scale and of routine proximity to other humans is affecting our families. Somewhere between 1945 and 2000, our idea of family and of this connectedness changed. With the exception of immigrant families, extended families today rarely live close together or in the same homes. Single occupant homes are far more prevalent than in the past. Certainly this is driven by preference, money, and life losses, but we should also be asking what the longer-term results might be of building so many places that isolate people rather than encouraging them to join in a supportive community.

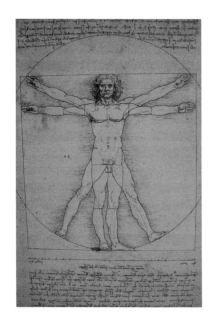

Figure 1.3 Leonardo da Vinci's *Vitruvian Man.*

Source: Flickr; photograph posted by Doug Williams.

Figure 1.4 Street scene in Greenwich Village in New York City.

Source: Photograph by Andrew Dalziel. Used with permission.

An important health experiment might be to build homes and communities meant to alleviate loneliness, not amplify it, to build homes and buildings that recognize the importance of extended families. (Plate 6 displays a map of a designed community that has taken the importance of proper social interaction into consideration.) Schools need to be in town, near where people work and where elder care facilities might be located, and not isolated in boxes on distant land. Children need older adults, and if they are fortunate, their grandparents, around them. In the pediatric clinic the nurses would observe that a good portion—although by no means all—of the questions they were asked could have been handled by a grandparent. Often a grandparent has the wisdom to listen and the time to spend with a child, much more so than a pressured parent. There is an African adage that "only the old can untie the knot." This generational connectedness is organic; it is built into our DNA, and we lose something when we let the habit of it go. Children need the elderly, and as anyone who has visited an elder care facility knows, elders need children around. The ideal home has private space for the family's elders. In Berkeley, these are called *in-law apartments*, a double meaning, I suppose.

Sometimes I think we are becoming increasingly indifferent to human suffering. My first thought is to blame technology, headsets, and game toys, or the isolation inherent in how we build our communities. Yet as I read through history, indifference seems latent in our DNA, like an oncogene (a cancer gene that lies dormant until stimulated by an environmental insult like radiation). When I was young, I almost never saw homeless people, and Newark was not a rich city. The mentally ill were washed, fed, and kept clean and safe in the large state hospitals. When I was a medical student, I did some training at the state mental hospital in Marlboro, New Jersey. It had a large, pleasant solarium; the rooms were orderly and clean. Yes, in many ways it was a warehouse, but now instead of this, we have desperately sad people on nearly every block in many U.S. cities. Psychotropic drugs, social work, and jobs were supposed to help these people, but the first thing mentally ill people need is a safe place to live. It need not be fancy, but it does need to be humane. The built environment comes first.

FOR LOVE OF COMMUNITY

What is it we love about a community? Maybe it is the weather—bundling up in winter or wearing shorts and flip-flops when it is warm, or hearing the rain against the window, or waking up to a fresh blanket of snow. Are we attracted to theater, movies, music, outdoor or organized sports? Maybe we are drawn to a community by specialized industry, opportunities for work, or the local schools, universities, and institutions?

Me? I really like feeling safe and being with people. I think this is not uncommon. I also like feeling snug from the elements. On the damp, windy, west coast of Ireland, a stranger can walk into a pub, listen to the locals talking, playing the bodhran or the penny whistle, or singing, and having a pint, and it is a sad fellow who does not feel happy. In fact the Irish often nickname the place a *snug*. Feeling safe and snug makes us glad, even in places as bleak and chilly as the west coast of Ireland in the winter.

Let's ask this from a different perspective. Do you like the way your built environment looks—the choice of building materials, the shapes and sizes of the buildings, the other architectural choices, the preservation of historic structures or the modern design of the landscape? Our communities, the ones we connect with, are each unique and known for their special mix, or omission, of characteristics. What would Boston be without the Common? What would New York City be without Central Park, Wall Street, or Yankee Stadium? (Central Park in the spring, shown in Plate 27, is a compelling example of an urban setting with unique drawing power.)

Our community creates a sense of both connection and stimulation in us. The connectedness comes from our memories of events that happened at a particular place. Driving through town, we find that places remind us of important moments in our lives. These buildings and places reflect a tapestry of stories from our past. The stimulation comes from what is actually happening in the environment while we are in it. There is energy in the flow of people and events.

One elected official I really admire is Mayor Joseph "Joe" Riley of Charleston, South Carolina. He understands the power of place, and he demonstrates the political leadership needed to create a vision as well. Mayor Riley worked with the community and developers to rejuvenate his city. Restoration and infill needed to be at the right scale in order to fit in as though it had always been there, only better. "It needs to fit into the neighborhood so it looks like it's part of the neighborhood's DNA," Riley says.

Charleston's design focused on the rhythm of the neighborhoods to create diverse, healthier built environments that minimize the conflict between cars and pedestrians and encourage people to have eye-to-eye contact. (Rainbow Row, displayed in Plate 8, is an eye-pleasing example of aesthetics and rhythm coming together in urban architecture.) Good design and great buildings do not just happen. They take political leadership, thoughtfulness, and architectural wisdom. The goal was simply to create a city that you would want to visit if you were setting off on an adventure or a vacation. In addition, as Riley says:

Every great city in the world, whether it's little or huge, has a healthy downtown. It's a public ground, it's a place that people own, where they celebrate their citizenship. Downtown Charleston is livelier and healthier than anytime in its history. It's filled with all kinds of people and all kinds of activity. People come from all over because there are very few places in our country where you can see anything that approaches the quality, diversity, excitement—and the joy of being on King Street.

Our community setting is the backdrop of our life, and as such, it can rejuvenate us. As we think about our daily movements or reflect on our commute to and from work, do we find we are having a calm, centering experience, do we gain energy from our surroundings, or do we find ourselves brimming with anxiety or even rage? Is there a favorite view, a noble tree, a beautiful statue, or piece of architecture to look forward to seeing each day, a scene that gives us serenity? Seeing such objects, enjoying their positions, and moving through them across the community build happiness.

June Williamson, coauthor with Ellen Dunham-Jones of *Retrofitting Suburbia* and professor of architecture at New York University, became fascinated by the topic of how to better measure happiness. These authors report that surveys of both Americans and the British show a steady decline in self-reported happiness from the 1950s to the present—a period that has also seen a consistent increase in affluence and wealth.[3]

The graph in Figure 1.5, depicting the relationship between reported income levels and happiness, supports Dunham-Jones and Williamson's argument and raises the question that if affluence does not increase happiness, what will? One of the most e-mailed articles in the *New York Times* during August 2010 was by Stephanie Rosenbloom. Titled "But Will It Make You Happy?" it describes how she rejected and limited acquisition in her life and how this lack of "stuff" reduced her financial stress and increased her overall sense of happiness.[4]

Rosenbloom echoes Dunham-Jones and Williamson, who add that those who study happiness often distinguish between the ways we make ourselves happy in the short term and long-lasting happiness. In other words, buying the big house makes us happy at first, but once we get used to it, ownership does not increase happiness. If the house is far from work, the daily

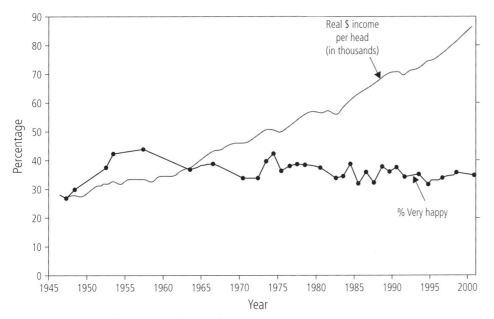

Figure 1.5   Income and happiness in the United States.

Source: R. Layard, "Happiness and Public Policy," *The Economic Journal* (Mar. 2006), *116*, C25. Used with permission of John Wiley & Sons, Inc.

commute is one more pressure on top of paying the mortgage, and we also miss the things we no longer have time for.

U.S. Census data for 1990 and 2000 show that most workers in the United States are still using a car, truck, or van for transportation to work (Figure 1.6). The most

dramatic shift is in the number of people working from home, which increased nearly 23 percent from 1990 to 2000.

According to these same Census data, 58 percent of workers commuted 24 minutes or less each way in 2000, down from 62 percent in 1990; however, there was a nearly 95 percent increase in the number of people commuting 90 minutes or more (Figure 1.7). Almost 8 percent of workers in 2000 commuted 60 minutes or more each way and accounted for 27 percent of the aggregate commute time. Fifteen percent of workers in 2000 commuted 45 minutes or more each way and accounted for 41 percent of the aggregate commute time.

Homes far from work and community mean we spend more of our lives in cars, cars that take us away from exercise and, just as often, from loved ones, healthy food, and safety. If we build schools too distant for walking, we isolate children, add to obesity, and take away from play and fitness. If we build streets and neighborhoods only for cars, we steal walking and socializing from our elderly and disabled residents. The more time we spend in a car, the more likely we are to be obese. The data on child obesity at the beginning of this chapter speak to families with little time. As I have mentioned, when I was a young pediatrician my colleagues and I never saw children with adult-onset diabetes. Now, between a third and a half of all new juvenile diabetics have what we call type 2 diabetes. And only 25 percent of this nation's kids can pass a very easy fitness test. If these trends continue, we are looking at the first

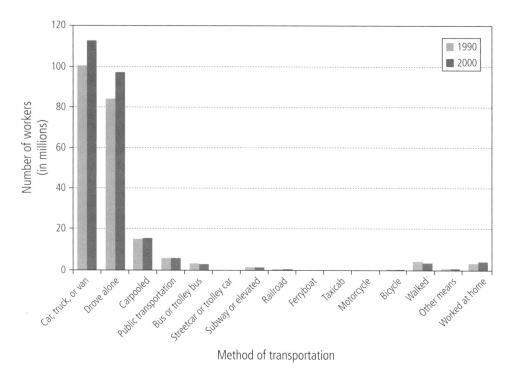

Figure 1.6 Means of transportation for U.S. workers aged 16 years and older.

Source: Data from the U.S. Census Bureau; graph by Simrin Cheema, UCLA School of Public Health.

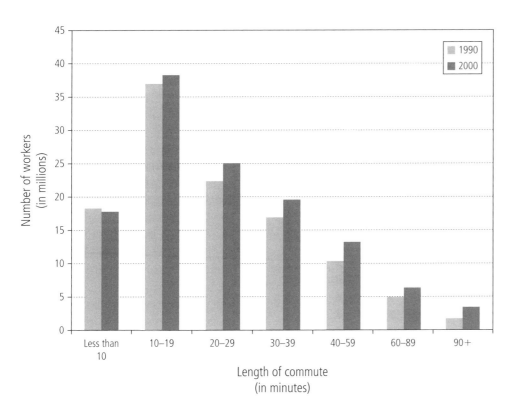

Figure 1.7 Travel time to work for U.S. workers aged 16 years and older.

Source: Data from the U.S. Census Bureau; graph by Simrin Cheema, UCLA School of Public Health.

Creating a Negative Place

If we set out to create a depression-inducing environment, the key features would include prevention of exercise, removal of interesting and pleasing natural features, excess noise, chaotically designed and located structures, and the creation of social isolation and anxiety—precisely what we built in the mid- to late twentieth century. (Plate 22 is an example of a development with such an environment.) As a result, we live in communities designed for automobiles and not for salving the human spirit. Americans spend many hours commuting long distances on high-speed, crowded roads. According to a 2007 Gallup poll, 38 percent of people traveling more than an hour to and from work said the commute was stressful. Sound systems and cell phones may mitigate the isolation of the modern automobile, but for most people, the commuting experience is rarely restful.[5] (Plate 26 exemplifies a typical stressful commute.)

According to the National Institute of Mental Health, major depressive disorder (MDD) is the leading cause of disability in the United States for ages fifteen to forty-four. MDD affects approximately 6.7 percent of the U.S. population, and although MDD can develop at any age, the median age at onset is thirty-two.[6]

The second most prescribed group of medications in the country is antidepressants. The costs of depression are substantial and include loss of lives, livelihoods, happiness, and productivity. Stress and anxiety often coexist with depression and amplify the disorder. The treatment

generation in history that will live less long than their parents did. Diabetes is dreadful. Unless it is very tightly controlled, it costs people their eyes, kidneys, feet, and eventually their lives. The best preventive treatment for type 2 diabetes is reducing the calories consumed and burning more calories off.

of mild to moderate depression involves medication and counseling, including cognitive therapy. Humans have had to cope with depression since long before these interventions were developed.

Nonclinical interventions for milder forms of depression relate very much to the built environment. The first is exercise, which increases serotonin levels and is documented as helping to alleviate moderate depression.[7] Contact and support from loved ones helps alleviate the acute depression associated, for example, with loss. Contact with nature and water has long been a mainstay for relieving *neurasthenia*, the historical term used to describe the condition of the chronically fatigued, wilted, frail, or depressed person of the 1800s.

Being Open to Cultural Shifts

Community is about people. We need to belong where we are, but we also need to have contact with new cultures and new stimuli and be open to them. I know this is classic American multiculturalism, but it doesn't make sense for us to be so diffused that we lose our identity or so hunkered down that we don't see a larger world. This is about finding balance in appreciating people's differences and commonalities.

As cultures shift, in part driven by economic changes, some areas decay as others improve. Although it is always good to have more services and quality recreation and food sources, formerly depressed communities are also confronting the threat of gentrification, a genuine threat to a city like Charleston. The city has become so desirable that it is pricing out the middle class and working people. Detroit is facing exactly the opposite set of challenges because it has deteriorated to such a degree that homes that were fabulous even thirty years ago are now gone, and the only people still there are those who can't afford to move. The city is shrinking before its citizens' eyes. (Plate 19 reveals an abandoned part of Detroit.)

In Central Detroit, the neighborhood decided to build a garden to grow food. What I love most about this idea is that it is not only about the growing of food; the act of building the garden itself grows community. (Plate 20 displays a typical community garden in Detroit.) And it grows children. I was watching the kids loading wheelbarrows and planting, and I thought, "What better kind of exercise could a child do than to be working with Mom, Dad, and the neighbors, planting in a garden?" It is a cultural shift for many folks. And given that eating the popular culture's high-fat, high-sugar, high-salt food is not just unhealthy, it is lethal, we would all do well to return to our farming roots. To have fresh fruits and vegetables, to grow them ourselves, to have a place of joy, a place of exercise, a place of fresh air—what could be better? The people in Detroit are creating urban farms where they can grow fresh foods, but what is more, they are using this change in their built environment to grow a connectedness to each other as they also grow healthier and stronger children.

Sharing Interests

There is something organic about Wy Livingston's Wystone's World Teas in the Belmar district of Lakewood, Colorado. It is not just that she sells organic food but that the whole process of choosing *place*, creating space, and building a community by being a part of a community is organic. Wy chose to open her tea shop in a town that once centered on a tumbled-down shopping mall with roving gangs. She was looking for a place that gave her a sense of belonging, like a nest—a nurturing environment.

When Wy was looking for a place that reflected her ideals and gave her energy, she wanted a place where people work, live, and play. She wanted a constant stream of customers—people who would stop by as part of their daily routine. She understood that communities need gathering places—coffee shops, tea houses, bars, whatever. (Plate 2 shows Belmar's central gathering area.) She wanted to be a part of people's lives on a personal and daily level. When she found the Belmar district, she found people who resonated with her business ideas and a culture that reflected her values and beliefs. "Colorado, in general, is very special—one of the healthiest states in the nation. As part of our brand we're obviously concerned with the environment, so coming to a place where people are very concerned about what they put in their bodies just became an objective in finding the right location. We're organic and Belmar generates some of their electricity through wind and solar. It's a good fit," she says.

Choosing a location for a new business should resonate at a deep level. We give to the community and in turn the community is an expression of what we value. Our values are woven into the matrix of people's lives. A community might be thought of as a series of concentric circles. We gravitate toward people and events based on common values and interests, and we may have many different interests and therefore different communities that we are a part of. There are the places where the artists hang out and also places popular with bikers, foodies, and sports fans. As an interest group takes hold, its members expand their influence over a block, a neighborhood, or a city. Sometimes these areas acquire names, like *restaurant row* or *the arts district*. Some cities reflect the local interests so keenly that we see, hear, and smell only them and become blind and deaf to other things. When we think of Paris, for example, we may think of fashion, paintings, and food, not of trash, noise, or crowds.

Sometimes these communities based on shared interests gather intermittently. People come from far and wide for science fiction conferences or political conventions. Other groups never actually meet physically at all. There are those who tweet, text, or participate in live-action video games, social networking sites, and other phone and Internet-based communities. I am not convinced these are communities. What I mean is that when you are isolated and you spend too much time in a virtual setting, it seems to me to be unfulfilling. I think you have to get out and see humans. It is not enough for us to live in our minds. We have bodies that need interaction with the tactile world. It all comes back to balance.

FOR LOVE OF OUR NATION AND THE WORLD

What does it mean to be where we are? Do we love the land (the mountains, plains, and deserts) or the idea of the United States (democracy and personal freedoms)? Americans have many identities, reflecting ethnic, religious, national, and personal interests. Part of what makes the United States unique is that its population is an *intentional* community. We have all, at some point in our personal or family history, *chosen* to be Americans, and during this history we have defined and redefined what it means to be an American.

Part of being an American, and maybe of our definition of citizenship, is having a sense of idealism—a hopeful view of the future and the belief that new directions will make our country stronger, better intellectually and economically, and kinder to its citizens. We have got to take care of ourselves, our families, and our extended circle, but we also have larger obligations. Our social frameworks of health care, public education, environmental protection, safety, and foreign policy are all a part of our idealism and our obligations.

I have said that to love is to want what is best for the beloved. If we love humanity, we must want people to be healthy. Being healthy includes not starving and not being subject to violence or preventable disease. In Russia, where I worked for the U.S. government on health and environmental issues during the Gore-Chernomyrdin negotiations, my team was looking at radiation hazards from both civilian and military sources. Using concerns about radioactive iodine from Chernobyl-like events, my colleagues and I were able to arrange U.S.-Russian collaborations that led to iodination of salt in Russia, Ukraine, and Belarus. The benefits of this intervention in terms of child health and survival were genuine. Sharing knowledge to benefit global health is in our interest. To take care of people elsewhere is to take care of ourselves in the long run. When we don't take care of the global community, that failure comes back to us in terrible ways.

Over my career I have worked in many roles: with community groups, schools, government, and hospitals. I hope I have contributed to the larger community, but I admit I have received more than I have given. Making places that build people, families, communities, and the nation is a rewarding way to spend one's life.

Chapter 2

What Is Health, and How Do We Measure It?

When we think about health, we often consider the very small and the very large, ranging from microscopic infectious organisms and toxic molecules to global climate change and the pandemic effects of war. Yet the major concern for human beings has been to be free from the tyranny of infectious disease.

Because infectious diseases, such as plague or tuberculosis, were so lethal throughout most of human history, their presence or absence has been a primary metric of health. Despite this importance, humankind was ignorant of the root causes of most infectious disease until the widespread acceptance of the findings of microbiology in the late nineteenth century, which resulted in what is generally known as *germ theory*.

As long as cities have existed on Earth, they have generally been squalid and dangerous places. Cities were frequently swept with epidemics of plague, smallpox, cholera, dysentery, tuberculosis, typhoid fever, influenza, yellow fever, and malaria. Rodents and insects thrived in these old cities and were the primary instruments that carried these diseases over the generations. These epidemics were accelerated by unsanitary practices such as the accumulation of waste, overcrowding, and a lack of clean water, food, and air.

For most people at most times, the leading overall causes of death were infectious and easily transmitted diseases. More than 30 percent of all deaths were among children under age five, with pneumonia, tuberculosis, and gastrointestinal infections the main causes of mortality.[1]

Figure 2.1 shows the ten leading causes of death expressed as a percentage of all deaths in 1900; Figure 2.2 shows the ten leading causes in 2006. The major causes of the decline in the diseases that were leading causes of death in 1900 were immunizations and better built environments. The discovery of penicillin in 1928, followed by other antibiotics, helped to bring down infectious disease death rates as well. Smallpox, a scourge for generations and a disease that decimated the American Indian population, has now been eradicated, thanks to worldwide "search and contain" immunization programs led by the World Health Organization

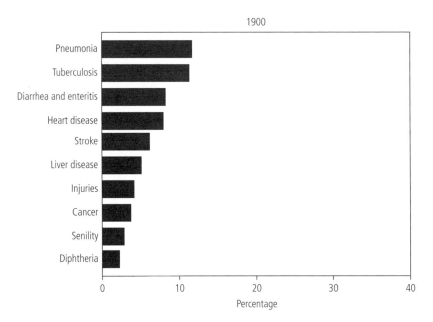

1900

Figure 2.1 The ten leading causes of death as a percentage of all deaths—United States, 1900.

Source: Centers for Disease Control and Prevention, "Achievements in Public Health, 1900–1999—Control of Infectious Disease," *Morbidity and Mortality Weekly Report* (July 30, 1999), *48*(29), 621–629, fig. 2.

fraction of their previous toll. Child mortality decreased, and average life span increased by over thirty years. Beginning in the year 1860, the average life span of a woman in the United States increased three months every year going forward, though this trend stopped about twenty years ago. Infectious diseases are in abeyance, but I and most other doctors anticipate their return, in part because of wasteful and indiscriminate use of antibiotics leading to resistant infectious organisms. When I was a young pediatrician, a resident once joked that penicillin was so effective at killing the *pneumococcus*, a common cause of pneumonia, that you could uncork the container on the other side of the room and the bacteria would be gone. Sadly, those days are long gone, and now powerful fourth-generation antibiotics requiring large-dose, in-hospital intravenous treatment are often required, and many physicians worry that it will not be long before even these powerful meds will become ineffective.

Of course there will always be a leading cause of death, as we all must die of some cause. The duty of scientists and physicians is to prevent *premature* death due to disease, because that has been the immediate challenge facing the human race since prehistory. Currently, infectious diseases have been replaced as leading causes of death by chronic and often environmentally induced causes—and I tend to think about environment writ large. This means that even though we have long, by and large, defined health as the absence of infectious or acute chronic disease, we must now revise and expand the definition of health to combat the causes of the causes of

(WHO), with essential support from the Centers for Disease Control and Prevention (CDC) and national programs. Polio, the reason that every year's class in my grammar school had a child with leg braces, and the disease that killed my young father, is close to eradication thanks to powerful CDC, WHO, and national immunization programs and astonishingly generous support from Rotary International.

In response to the advances in biological sciences during the nineteenth and twentieth centuries, death rates due to preventable infectious diseases dropped to a

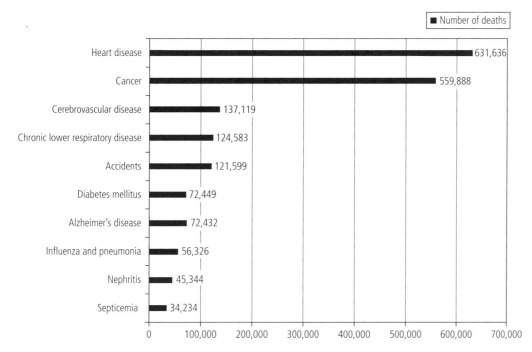

Figure 2.2 The ten leading causes of death as a percentage of all deaths—United States, 2006.

Source: Data from M. Heron, D. Hoyert, H. Murphy, and J. Xu, "Deaths: Final Data for 2006," *National Vital Statistics Reports* (April 17, 2009), 57(14).

death—causes that take lives prematurely today and that diminish the quality of life for millions of other people.

PERSONAL HEALTH

Today the primary enemies of good health are those that we as a society inflict on ourselves—an unhealthy environment, lack of exercise, and bad nutrition. As we saw in Figure 2.2, seven of the leading causes of death in 2006 were heart disease, cancer, stroke, chronic lung disease, unintentional injury, diabetes, and pneumonia and influenza. The common thread in this list of maladies is their relationship to the environment in which we live and the lifestyles we choose to follow.

10,000 Steps

In 1965, the Japanese introduced a pedometer called the *Manpo Kei*, which translates as "10,000 Steps Meter." This name is derived from the belief that taking 10,000 steps each day greatly enhances one's personal health through exercise. In studies of over 3,000 people with pre-diabetes in the United States, those who walked or exercised five times a week for at least thirty minutes reduced their body weight by 5 to 7 percent and reduced their risk of diabetes by 58 percent.[2]

Walkable neighborhoods encourage a 10,000-step lifestyle. A 2003 article in the *American Journal of Health Promotion* noted that "older women who live within walking distance of trails, parks, or stores recorded significantly higher pedometer readings than women who did not. The more destinations that were close by, the more they walked."[3]

There are billions of people alive today, including the millions in the United States, who are living in the midst of pollution, poisons, and sources of infection at a very personal level. This is not only the situation of the impoverished. The immediate environments of home and work, even for middle- and high-income individuals across the United States, are taking a toll on personal health.

The Residential Microenvironment

The immediate built environment in which we live and work has a profound effect on our personal health. Most of us recognize a threat instantly when we smell smoke or gasoline in our home or workplace, but other health factors are more subtle and harder to pin down yet just as dangerous over time. For example, at least 70 percent of cancers are due to the environment.[4] As those in environmental health say, "The genes load the gun. The environment pulls the trigger." The physical location or positioning of our homes, schools, or workplaces can expose us to the harmful effects of natural phenomena. For example, radon, a radioactive gas that forms through the natural decay of uranium, is the leading cause of lung cancer among nonsmokers and is the second leading cause of lung cancer in the United States. The U.S. Environmental Protection Agency (EPA) estimates that radon is the cause for 20,000 lung cancer deaths annually. Granite contains uranium that leads to the release of radon, an odorless gas that our bodily senses cannot detect. Radon released from soil containing granite can enter a structure, like a home, through cracks in basement floors and walls or via gaps around underground pipes as they enter the building. Sealing these cracks and openings and maintaining air circulation can mitigate many potential problems.

The construction of homes, schools, or workplaces can be a source of chronic allergies and asthma or irritations caused by volatile organic compounds (VOCs) or natural agents such as dust mites, cockroach feces, and spores from fungi. Balance in an ecosystem depends upon the decomposing of organic material by fungi and molds so that nutrients can be reused. Although molds, such as black mold, reproduce naturally in areas with high humidity, a building that is excessively wet during construction or is not sealed properly can become a long-term mold incubator. Unattended leaks and water intrusion after construction are not just cosmetic and economic problems; they are health challenges as well.

Choices made in home construction materials and furnishings can also affect our health over time. Many houses built before 1978 and those furnished with old furniture likely contain lead-based paint. If this paint peels, chips, or cracks, it can be hazardous. In such dwellings the dust around windowsills and on friction surfaces like doorjambs can contain highly bioavailable lead. The soils around older houses can retain lead deposited from the flaking and "chalking" of exterior lead-based paint or from leaded-gas car exhaust. Lead can enter drinking water through old lead pipes and from lead-based solder. Lead poisoning has been linked to many growth, reproduction, movement, and nervous system operation impairments. A child eating a single lead paint chip can become lead poisoned.

When I was growing up, paint was 50 percent lead; in fact the weight of a paint can was largely that of its lead content. In those days the definition of lead poisoning was a blood-lead level greater than 60 micrograms (mcg) per deciliter. Mine was about average, about 22 mcg per deciliter. Now we know that no level of lead is healthy in the body, especially in the brain. But lead was everywhere not too long ago: in our water, paint, canned food (from the metal solder on the top of a food can), and very important, in gasoline.

Lead was used in printing for hundreds of years, and back in the 1750s Benjamin Franklin described printers developing the "dangles," that is, a *wrist drop*, and losing the ability to move their hands properly because of lead toxicity.[5] One hundred and fifty years ago workers in lead foundries were known to develop serious illnesses, including one called Saturnine gout, after the ancient association of the planet Saturn with lead. It was well known that women who worked in lead foundries were much more likely than other women to have miscarriages, babies with birth defects, and babies that did not grow properly.

It was not until the late 1970s, thanks to the work of a heroic pediatrician and psychiatrist, Herbert Needleman, that we began to realize that lead had effects far longer lasting than immediate toxicity. In a remarkable study, he collected the shed baby teeth of six- and seven-year-olds

(he paid for them). At the same time, he collected psychological measures and performance assessments by the children's teachers. The teeth were important because they reflected each child's lead level before age two, back when the teeth were being formed and about the same time that the brain was undergoing very rapid growth. Dr. Needleman showed that the higher a child's lead level, the more likely the child was to be hyperactive and distractible and to have learning problems and poor academic performance. This and subsequent studies had major influences on our thinking about low-dose exposures to chemicals, particularly in susceptible populations such as young children.

Reducing blood lead levels by 10 mcg per deciliter raises average IQ in a population by an average of 2.6 points (1.9 to 3.2).[6] We can debate the value of IQ tests, but they are pretty good predictors of lifetime income. In economic terms, the value of an IQ point is estimated to be about $14,500 (in year 2000 dollars) of lifetime income. An article I coauthored for *Environmental Health Perspectives* in 2002 discussed these findings and calculated that the substantial reductions in lead exposure among the children of the United States has saved each year's cohort of children between $110 billion and $318 billion per year in lost economic potential and increased health costs.[7] When I was at the Centers for Disease Control and Prevention we had a goal of eradicating lead poisoning by the year 2000. We weren't completely successful, but now average blood-lead levels in this country are less than 2 mcg per deciliter.[8]

Figure 2.3 Asbestos fibers.
Source: Photograph from the U.S. Geological Survey.

Asbestos is a fire-resistant fiber used in construction materials including roofing shingles and ceiling and floor tiles (Figure 2.3). When the asbestos is in an undisturbed condition, it causes no threat, but when fibers of asbestos become airborne, exposure to them can cause lung cancer and *mesothelioma*, the dreadful cancer of the linings of the lung and abdomen. These cancers kill thousands of people per year. I, and many of my colleagues in environmental public health, feel the use of asbestos should be banned worldwide.

Carpeting can be the cause of chronic respiratory illness, headaches, fatigue, and skin irritations, either through the off-gassing of VOCs from new carpet or from the decaying material, dusts, and allergens that accumulate in carpeting and the padding below it.[9] When working in an allergy clinic, I was taught that carpeting supports one thousand times the allergens that a bare or tile floor does. Doctors routinely recommend to families of children with asthma that the carpets be removed from the house where possible, and that the floors be kept bare. Additionally, carpet is currently 1 to 2 percent of all the solid waste in our landfills. There are companies that are now picking up old carpets and recycling the plastic fibers. This is a model for the future.

When we look at our built environments, our goal should be to put a solution in place that mitigates multiple problems, yet we cannot legislate people's choices in personal health. For ourselves, although we may not know the original construction of our homes, schools, and workplaces, we can choose or advocate for wall and floor coverings not associated with health problems, as well as choose whether to open a window for fresh air, and select appropriate plants when landscaping.

When we make places, we are making health or disease. The greatest impacts on health do not come from medical care but from *upstream determinants*. This is

Walk to School Like Grandpa Did

In 1974, 66 percent of all children walked or rode a bicycle to school. By 2000, that number had dropped to 13 percent, and childhood obesity had skyrocketed. By 2004, just 25 percent of California's fifth graders passed the annual fitness tests and the cost of transporting California's children to school by diesel bus had climbed to more than $1 billion per year.

Studies have shown that walking or bicycling to school increases children's concentration, improves mood and alertness, and enhances memory, creativity, and overall learning. Programs that promote safe routes to school for children result in improvements in both academics and physical fitness. When infrastructure and social programs create and support those safe routes, schools in areas with initially low levels of walking or biking to school show increases in these healthy behaviors by 20 to 200 percent.[10]

where we get the most change for good or ill with the fewest costs. A major upstream determinant is how we build where we live. This affects how much we eat, walk, and socialize; the safety of our home and community; and the quality of the air we breathe and the water we drink.

The Obesity Challenge

Many unhealthy factors actually stem from our habits and not from any deficiency in our environment. How many homes in the United States have exercise equipment that sits unused while the occupants lounge in front of the television?

In generations past, physical activity was woven into all the activities of daily life. Few in the United States were obese and most would never imagine joining a fitness club just to go out and exercise on machines. Fitness was taken somewhat for granted, and many working people longed to be free of backbreaking labor. As a nation, we got our wish in the second half of the twentieth century, but the unintended consequences have been devastating (see Figure 2.4).

Figure 2.4 Obesity and physical inactivity in the United States, 2005.

Source: Self-reported data from the Behavioral Risk Factor Surveillance System, Centers for Disease Control and Prevention, http://apps.nccd.cdc.gov/gisbrfss/default.aspx.

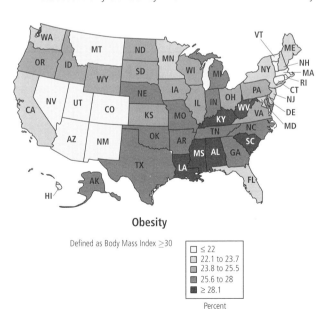

Obesity

Defined as Body Mass Index ≥30

- ☐ ≤ 22
- 22.1 to 23.7
- 23.8 to 25.5
- 25.6 to 28
- ■ ≥ 28.1

Percent

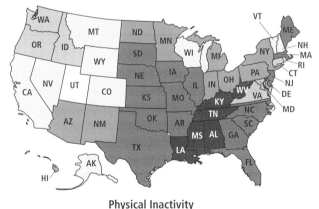

Physical Inactivity

Defined as adults reporting "No"
- ≥30 min moderate physical activity ≥ 5 days/week, OR
- ≥20 min vigorous physical activity ≥3 days/week

- ☐ ≤ 45.8
- 45.9 to 49.3
- 49.4 to 52.4
- 52.5 to 54.8
- ■ ≥ 54.9

Percent

Children have gained a dozen pounds in less than a generation. Three-quarters of high school students cannot run or walk a twelve-minute mile. Obesity was recently named the leading chronic disease health threat in the United States. Since 1980, the number of obese children in the United States has tripled, and among adults, obesity now rivals smoking as the largest cause of preventable death and disease. (See Plate 1 for the state-by-state prevalence of obesity.) The cost of obesity-related illness has been estimated at between $147 billion per year[11] and $215 billion per year when direct medical, productivity, transportation, and human capital costs are examined,[12] far exceeding all the costs for cancer.

In response to the increasing epidemic of obesity in the United States, First Lady Michelle Obama recently launched a campaign called "Let's Move: America's Move to Raise a Healthier Generation of Kids." The campaign aims to foster collaboration among the leaders in government, medicine, athletics, science, business, and education to encourage, support, and pursue solutions tailored to those children and families facing a wide range of challenges and life experiences. This includes empowering consumers with health-related information, encouraging people to monitor their weight and understand healthy parameters, providing more nutritious school lunch choices to build healthy eating habits, opening farmers' markets with fresh and locally grown foods in communities that lack ready access to fresh produce, and championing physical activity and fitness.

The built environment drives driving, and driving helps to drive obesity. Areas that have higher densities of fast-food stores have higher rates of obesity. Children whose schools are close to fast-food stores are more likely to eat high-calorie and low-nutrient foods, which promote obesity. According to a 2004 article in the *American Journal of Preventive Medicine*, predominantly African American and low-income neighborhoods average 2.4 fast-food outlets per square mile, whereas more affluent neighborhoods average just 1.5 fast-food stores per square mile.[13]

Those who live in low-density suburban and rural areas are more likely than inhabitants of higher-density areas to drive cars everywhere they go.[14] Both children and adults who live in areas without good walking or bicycling amenities are virtually forced to use cars by the design of the built environment around them.

Conversely, changes in the built environment can positively affect the rate of obesity. People who live in more dense areas use mass transit more often and are more physically active than those in suburban environments. Those who live near parks and destinations are more likely to walk or bicycle. Children who live close to schools are more likely to bicycle or walk to school. Persons who live near green space and more attractive environments are more likely to use physically active transportation.[15] Those who live near a neighborhood farmers' market are more likely to eat fresh fruits and vegetables.

One proof of the potential beneficial effect of the built environment is this: living in areas with walkable green spaces positively influences the longevity of urban senior citizens independent of their age, sex, marital status, baseline functional status, and socioeconomic status.[16]

Consider two average suburban men. Both are fifty-two years old, work in standard office jobs, and complain of low energy and general lethargy. This is not an uncommon medical scenario.

One of these men commutes twenty-five miles each way to work, spending three hours a day in his car. A physical examination comes up mostly normal, except that he's twenty-eight pounds overweight. His blood pressure is 155/95, and his blood sugars are elevated. He's starting to show signs of depression. His doctor treats him in the usual American way by scheduling a meeting with a nutritionist and recommending a weight loss program. The man buys a membership to a fitness center and generally tries to "get more in control of work and life commitments." As anyone might guess, within several months the exercise program has gone nowhere because he has no time to get to the fitness center and no good place to walk in his community or at work. He is now on blood pressure, blood sugar, and cholesterol medications and an antidepressant. His prescription medication costs run approximately $385 per month.

The second man has about the same physical profile, but after his checkup he changes his habits. He now bicycles just thirty minutes to and from the transit center three days a week. Each thirty-minute bicycle trip burns 250 calories in his 190-pound body, which amounts to 1,500 calories a week, and 78,000 calories per year. That works out to twenty-two pounds of body fat lost if he changes nothing else in his lifestyle. After one year, he has reduced his weight to 168 pounds, his blood pressure is 130/78, his blood sugars and cholesterol are normal, and he finds his energy and mood are both much improved.

So, what's the difference? Part of the story is the environment. The first man has a long commute and no place to easily get physical activity. The second man has an environment that supports healthier choices.

Our behaviors are not only affected by our physical culture but influenced by our social culture as well. When our neighbors are healthier, we are healthier. That said, the healthiest environment does not guarantee that we will *be* healthy. We have to decide to step up and take advantage of what is around us. Good neighbors can even help.

PUBLIC HEALTH POLICY

Another driving force of change is the influence of policy. The purpose of public health is to ensure the conditions where people can be healthy.

Diseases spring from the environment when the conditions are optimal. For example, plague has been present throughout recorded history and still exists. In the middle ages, more than a third of Europe's population died from the plague bacterium, spread by infected fleas and rats.

Even now, there are plague outbreaks in North America, Africa, Asia, and South America, claiming 1,000 to 3,000 lives each year.[17]

The movement of people can hasten the spread of disease. Consider the U.S. Civil War. Recruits often came from farms and small towns where they were in close contact with perhaps ten to fifty neighbors. They joined the armed forces and arrived in squalid camps with hundreds or thousands of fellow soldiers. These young men were in unfamiliar and cramped settings where they were exposed to germs for which they lacked immunity. Many never made it out of those camps to fight, as they got sick and died from infections that spread from soldier to soldier. During the Civil War, battlefield deaths amounted to over 200,000. During the same period there were 400,000 deaths from disease.

As I recounted in the Prologue, Frederick Law Olmsted was not only the architect of New York City's Central Park but also Abraham Lincoln's appointee to head the U.S. Sanitary Commission during the Civil War. The work of the commission doubled the survival rate of injured Union soldiers. Rather than worry about specific diseases, Olmsted oversaw changes that made sure that the hospitals—the dressings, food, and water—were clean. He made sure that sick soldiers received sunlight and fresh air. He put sanitation into hospitals before the widespread acceptance of the germ theory of disease.

What Olmsted understood at a very fundamental level is that science can tell us what to do to inhibit the spread of disease, but all three prongs of public health

must be put in place to effectively change the environment: *science*, *finances*, and *marketing*. Science tells us what the problem is, the budget tells us how to fund the changes needed, and marketing uses communication tools to influence public behaviors. Public health must be interdisciplinary. It must also have critical links to the medical care world to understand what causes infectious disease.

When I was California State Health Officer, the hardest battle I fought was not over obesity or disasters. It was over tobacco and particularly antitobacco advertising. Every time the state would raise the tax on tobacco, the tobacco industry wanted to use the money from that tax to fund care for those with lung diseases. What the tobacco industry did not want was for money to be spent on preventing people from smoking.

California has done better than the rest of the country at reducing smoking rates, going from 123.3 packs per capita per year in 1987 down to 43 packs in 2006 (Figure 2.5). The rest of the country still smokes about 90 packs per capita.[18] The interventions that were put in place in California included taxes on tobacco, used in part to fund the antitobacco program, antitobacco advertising, and environmental tobacco smoke laws. Messages were and continue to be communicated through public education, including public television. Then the PBS station in San Diego focused media messages on the dangers and consequences of smoking, and that worked. Twenty years ago, California's lung cancer death rate was somewhat higher than the rate for the rest of the

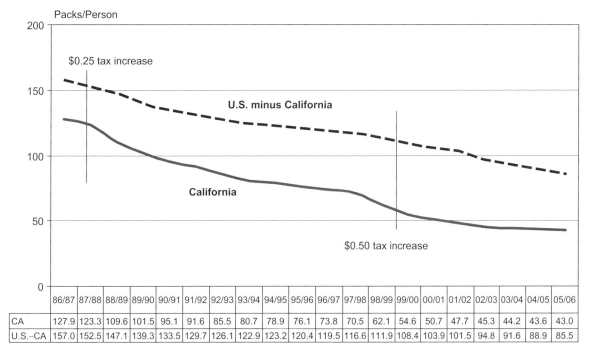

Packs/Person

$0.25 tax increase

U.S. minus California

California

$0.50 tax increase

	86/87	87/88	88/89	89/90	90/91	91/92	92/93	93/94	94/95	95/96	96/97	97/98	98/99	99/00	00/01	01/02	02/03	03/04	04/05	05/06
CA	127.9	123.3	109.6	101.5	95.1	91.6	85.5	80.7	78.9	76.1	73.8	70.5	62.1	54.6	50.7	47.7	45.3	44.2	43.6	43.0
U.S.–CA	157.0	152.5	147.1	139.3	133.5	129.7	126.1	122.9	123.2	120.4	119.5	116.6	111.9	108.4	103.9	101.5	94.8	91.6	88.9	85.5

Figure 2.5 California and United States minus California adult per capita cigarette pack consumption, 1986/1987–2005/2006.

Note: Dates are fiscal years (July 1 to June 30).

Source: Prepared from state and federal data by the California Department of Health Services, Tobacco Control Section, Mar. 2007.

country—today it is more than 20 percent lower. When we want to change people's behavior, simply telling people what to do or not to do does not work well. The California antitobacco ads had to find a message that made people angry. The most effective ad was the one with tobacco executives laughing about how many kids they had addicted that day. This "I am being exploited" message resonated with young people.

California also had to create conditions that made people want to be healthy. If I want to change behavior, it is most effective to change the environment and let that environment influence behavior. It was very important that California put in place environmental tobacco smoke laws that reduced and mostly eliminated smoking in buildings.

Public health policy does not only react to outbreaks of disease; it must also forecast epidemics and act proactively to protect the population. Through regulation and implementation of new technologies, people are relatively confident that they can go into a building, eat at a restaurant, drink water from a tap, or go to work without being exposed to life-threatening diseases.

Nevertheless, sometimes there are surprises. In 1976, American Legionnaires attending a convention in Philadelphia developed an unexplained and often lethal pneumonia. I was a CDC Epidemic Intelligence Service officer at the time and had the opportunity to assist in this outbreak. It took months of investigation, physical examinations of the sick and well legionnaires, and sampling from the environment before the cause was identified. Dr. Joseph McDade of the CDC identified some novel organisms in the lung tissue of legionnaires who had died. Further investigation discovered that the water in the air-handling system had become extensively contaminated with what was later called the *Legionella* organism, while the disease became known as Legionnaires' disease. As a result of these findings, substantial changes were required in the design and inspection of air-handling systems, with particular attention to preventing water from infiltrating the system.

Coincidentally, when I returned to the CDC in 1994, almost twenty years later, one of the outbreaks being wrapped up at the time was a *Legionella* epidemic associated with cruise ships. The CDC eventually traced the illnesses to whirlpool spas on the ships' decks. When the environmental specialists examined the filters for these hot tubs, they discovered that the filters had not been maintained or cleaned. As a result, regulations and design changes were put in place to both enable better record keeping and to require easy and regular changing of filters.

There is a larger lesson to be gained from these responses to outbreaks of illness: we can learn a great deal from such *natural experiments*, including things we need to do elsewhere to protect people. Additionally, they remind us that every new technology, even when it is thought to bring a basketful of benefits, is always accompanied by impacts on health.

We are also surprised when people get sick from food; we expect our food to be safe. Recall the *E. coli* outbreak linked to spinach that killed three people and left over two hundred sick in February 2009, or the outbreak of illness across the Northeastern states from romaine lettuce in May 2010. It is remarkable how much more common and visible, as well as genuinely immense, outbreaks of illness caused by unsafe food have become. Over the recent decades, food production has become substantially centralized and profoundly intermingled. This means, for example, that even if a relatively small plant is supplying peanuts that turn out to be contaminated with *Salmonella*, those peanuts can, over a short period of time, contaminate hundreds of food products, ranging from peanut butter to candy bars, and can cause illness in thousands of people and in nearly every state.

One day in 1999, I was driving along Buford Highway in Atlanta to CDC headquarters for a meeting. The road (it is really not a highway) is seven lanes wide—three in

Figure 2.6 The dangers faced in crossing Buford Highway point to upstream issues.
Source: Photograph from the Media Policy Center.

each direction and a lane for right or left turns (a "suicide lane") down the middle (Figure 2.6). The temperature was 95 degrees, and the humidity was 95 percent. I looked to the right side of the road and saw an elderly woman struggling along with plastic shopping bags in each hand. She was bent over with osteoporosis. She had red hair and looked like my mother.

I went to the meeting to discuss national and global health issues including causes of death, but all I could really think of was that poor woman. If she were to collapse, the cause of death would be listed as heat stroke, and there would be no mention of the upstream causes—the absence of trees, a black tar road, too many cars, too much air pollution, and lack of public transportation. If she were killed by a truck, the cause

of death would be listed as motor vehicle trauma, and again, the real causes of death, the lack of sidewalks, lack of public transportation, poor urban planning, and failed political leadership, would be misdiagnosed or, worse, ignored.

Part of the work of public health is to examine the causes of death. When I and others in public health look upstream at the real causes of morbidity and mortality, we can make discoveries. First, we can recognize that health professionals have very little involvement with or effect on the *real causes* of death, even though I think they should. Second, we can see that these real causes of death are being driven by social, ideological, and economic realities. And third, we are likely to learn that when public health doctors begin to focus on upstream causes, it can generate a powerful and unwanted political backlash.

Nevertheless, public health policy needs to look at these macro-issues affecting the health of the population, and it also needs to consider micro-issues, problems that can grow undetected long enough to infect our food, water, air, and mechanical systems. At this microlevel, public health policy needs to consider how what happens to individuals influences the greater community. Poverty, lack of education, inadequate mass transit, crowding, and a variety of other negative factors affect not only the individuals directly involved but also the surrounding community. Communities can look at their local environment in terms of the potential for health concerns, and can come to understand that public health is about providing the conditions necessary for people to live a healthy life.

Buford Highway, for example, where I saw the woman with the shopping bags, is less than a mile from and parallel to Interstate 85. It absolutely does not need to be seven lanes wide or to have a mile or more between crosswalks. A concerned community must examine why this woman must carry heavy packages alone and such a far distance. It needs to examine how public transportation can support her needs. And what is in those packages? Has she purchased healthy food or highly processed meals that are easy to prepare? It should examine those services that can help provide nutritious, low-salt, and low-fat meals to the elderly. Public health becomes very personal when we center our thinking on an individual and how we can bring resources to bear to make it possible for that person to lead a healthier life.

ENVIRONMENTAL HEALTH

Our built environment is so pervasive and we human beings are so adaptable that we hardly notice it. In fact, the profession of environmental health scarcely paid attention to the built environment until the second half of the twentieth century. *What* we build and *how* we build it affects our overall environment and thus also affects the viability of our children's future.

In the nineteenth century the United States improved the built environment by removing biological and industrial waste and by supplying clean water, safe food, home heat and lighting, and mass and personal transit. Construction reduced crowding, and developed healthy cities and trolley car suburbs. In the twentieth century, thanks to cheap and abundant fossil fuel and unparalleled prosperity, the United States was rebuilt for automobiles. As populations grew, small towns became cities, and farmlands and forests were turned into subdivisions, industrial parks, and highways. Public transit systems like trolley cars were abandoned; schools became larger and more distant from towns, commutes longer, and traffic congestion overwhelming. Fresh water became increasingly more precious, and the atmosphere hotter and more polluted. We are only now beginning to understand the toll of these choices on our bodies.

Deep in the American psyche there is a feeling that when we have destroyed the place where we are, we can always go someplace else, yet the reality is that there are scarcely any *someplace elses* left. We have to take care of where we are.

Through travel and technology we can now transit the globe in days rather than months or years. What happens halfway around the world can affect us just as acutely as events that occur close to home. Witness the radiation scare resulting from earthquake damage in Japan that we are experiencing as this book is being written. Due to the natural movement of air in the Earth's atmosphere, particulates can move globally. What happens in Europe or Asia can affect air quality in the United States, and vice versa.

Recently, an international consortium of scientists used satellite data and computer modeling to determine the impact of the Asian monsoon season on global air circulation. Examining the upward flow of air from Asia, they learned that this air, carrying a variety of pollutants, ascended into the stratosphere, where the pollutants traveled the globe for several years, eventually descending back to the lower atmosphere or breaking apart. It is suggested that the level of pollutants entering the stratosphere will grow with the increase in industrial activity in rapidly developing nations, such as China. Why is this important? The compounds in the pollutants become an aerosol that is known to affect the ozone layer in ways that reduce our planet's protection from ultraviolet radiation, which causes skin cancer. Hydrogen cyanide, produced largely from burning trees and other vegetation, is recorded at low levels over the tropical ocean but at high levels over areas in Asia where land has been recently cleared for agriculture. This and other pollutants in the stratosphere have been linked to an effect on levels of water vapor that in turn increases the capture of solar heat on the Earth, affecting the global climate.[19]

Precipitation is naturally a little acidic due to chemical reactions in the atmosphere, but this natural acidity can be increased by human-induced activities. In the United States, nearly two-thirds of sulfur dioxide (SO_2) emissions and about one-fifth of nitrogen oxide (NO_x) emissions are produced by the burning of fossil fuels by electricity plants. Nearly one-half of NO_x emissions are from sources of transportation.

Prevailing winds transport these gases over long distances, crossing state, national, and international borders. Eventually they combine with moisture to fall as acid rain, which accumulates in ponds and lakes, and on soil. In time the slightly acidic water can dissolve and carry along metals from the soil into water sources, ultimately causing chronic stress in animals in the area.

What does acid rain have to do with the built environment? Plenty—one-third of all the fossil fuel we consume goes to the heating and lighting of buildings and another third goes to our vehicles, to move us to and from the built environments we construct. The air pollutants that produce acid rain also affect us directly. As we interact with the atmosphere, fine sulfate and nitrate particles are inhaled deep into our lungs, increasing illness and premature death from heart disease and lung disorders, including bronchitis and asthma.

The question is not whether there will be pollution, but what we are going to do about it. It is up to us to change how we approach pollution. In the United States we have the Clean Water Act and the Clean Air Act, two important pieces of legislation that have made significant changes in how we do business. Through the 1960s, there were few if any pollution controls. The original passage of the Clean Air Act in 1970 enacted a major shift, authorizing comprehensive federal and state regulations to limit emissions from both stationary (industrial) sources and mobile sources. There have also been amendments to the original legislation (in 1977 and 1990), each authorizing tighter regulation of emissions and setting higher safety standards.

The greatest intervention for reducing levels of atmospheric lead in the United States occurred in Los Angeles. The federal government gave California a special dispensation for stricter emission controls than those then mandated for the nation. The tighter regulations sparked technological breakthroughs. Beginning in the 1970s, cars became equipped with exhaust gas recirculation pumps and catalytic converters, positive crankcase ventilation, and evaporative emissions canisters for fuel tank fumes. Leaded gasoline destroys a car's catalytic converter. Today our cars have computerized engine management, better catalysts, and use unleaded gas. The engines in the latest generation of diesel cars consume about 30 percent less fuel than gasoline engines of equivalent power. Using ammonia-based acid injection into the exhaust stream, these diesel cars emit about 25 percent less carbon dioxide and 90 percent less nitrous oxide than gasoline-powered automobiles. Particulate emissions have been reduced by 98 percent compared to the diesels of the 1970s. So, have these changes had an impact on the quality of our air?

According to the EPA, "today's cars emit 75 to 90 percent less pollution (for each mile driven) than their 1970 counterparts,"[20] owing to improvements in vehicle and fuel technology and increasingly stringent standards for tailpipe emissions. Yet even though emissions from cars have dramatically decreased, our air quality overall has improved only incrementally and not nearly as dramatically as the reduction in emissions. Why? One reason could be that midsized and larger sport utility vehicles (SUVs) are classified as trucks, not cars, so they do not have to follow the most stringent emissions and fuel economy standards. With the increased popularity of SUVs, there are more vehicles on the road with lower standards for emissions, reducing the potential benefit emission regulations hoped for. Heavy trucks, jets, and other large vehicles also contribute to air pollution. As of April 2010, tighter emissions standards for new cars and trucks were approved for fleets of vehicles, beginning in model year 2016.[21]

Another reason could be that even though emissions per car have decreased by more than 70 percent since I got my driver's license, the number of drivers has increased three- to fourfold. All the benefits we could have had have been washed away by not investing in public transit and quality density. So the net effect of emission controls on passenger cars is less than it should be. What we can see from this example is that where regulations are in place, change happens, but other factors often come into play, so evaluating improvements is essential—midcourse adjustments are nearly always required.

MENTAL AND SOCIAL HEALTH

People underestimate how extraordinarily important mental health is. Health is about our mental, physical, and spiritual well-being, not just the absence of disease. This isn't a novel idea, but one proposed in the constitution of the World Health Organization, dating from 1948, which states that "health is a state of complete physical, mental and social well-being and not merely the absence of disease or infirmity."[22]

Mental health is inherently biochemical and can affect our physical health by influencing how we feel about ourselves and the condition of our life. The immediate status of our mental health affects the choices we make every day and can be influenced by neurological imbalances, congenital disorders, or diseases that require medical care; by uncontrollable events that affect or change our life circumstance; and by our connection to others and the space in which we live. Our mental health status is something we feel personally, but this status also affects those around us at work and at home. In time, our personal mental health can affect our community, and the community can, in turn, affect our personal mental health.

Our health stems from the interrelationship of our personal, family, work, and community factors and our mental health. It can be influenced by world events, the weather, personal trauma, interactions with others, and nature. These are influences enough, but the one influence that we have a good chance of doing something about is our built environment.

When I was in Elgin, Illinois, preparing for this book and the video series, I heard high-schooler Ashley Lundgren talk about attention deficit disorder (ADD) and attention deficit/hyperactive disorder (ADHD). Her research showed that up to 10 percent of students are being diagnosed with ADD or ADHD and approximately 35 percent of high school dropouts are students with ADD or ADHD. She found studies that say students are better able to concentrate when they are in contact with nature. So, what about bringing nature to the city? Community gardens bring people together where they can interact with their neighbors, breaking down social isolation. Isolation and alienation are big factors to address in depression and stress management.

In 2003, researchers found that American children were twice as likely to be prescribed antidepressants at that time than they were five years before.[23] The highest increase (66 percent) was in preschool children. Why would a preschooler need an antidepressant? Mental health and stability are influenced by nature and green spaces. People are happy when they are outside. They're happy when they're in nature.

When you look at the happiness of a country or groups of people, people are happy when they make more money, up to the point where their needs are met. Once their needs are met, having more money does not make them measurably *more* happy. Data from the National Opinion Research Center's General Social Survey, when analyzed by researchers at Harvard University, revealed that people who feel connected to others, and who trust those they spend time with, actually live longer.[24]

This is not a new idea. In the 1970s, the Prince of Bhutan rejected the assumption that a country's gross domestic product is the appropriate indicator of a country's well-being. King Jigme Singye Wangchuck decided to make his country's priority *gross national happiness*.[25] His focus was to ensure that prosperity was shared across society and that there was a balance among preserving cultural traditions, protecting the environment, and maintaining a responsive government. The government of Bhutan has made big decisions based on this philosophy.

In September 2009, the president of France, Nicolas Sarkozy, suggested the development of a "happiness index," as the "dogma of economic growth is no longer sustainable. It has to be broadened into a new measure of political success and national achievement that takes account of the quality of ordinary lives and our professed desire to save the planet from environmental disaster."[26] This *net national product* indicator is still in development. Even our Declaration of Independence includes the "pursuit of happiness" as an "unalienable right," so why don't we measure it? Measuring a society only by dollars and not by other kinds of capital is a mistake.

So, in a book about the built environment, why is mental health an important topic? As a society, we all pay the price for those with poor mental health. Mental illness takes a deep toll on the immediate family, and on society when communities must provide services and manage care when situations rise to the level of crisis. When people are stressed because they cannot find work or are unhappy in their relationships, they are unable to contribute to the community as a whole and sometimes they vent their frustrations in violent or self-destructive ways. As James Howard Kunstler has written:

> We drive up and down the gruesome, tragic suburban boulevards of commerce, and we are overwhelmed at the fantastic, awesome, stupefying ugliness of absolutely everything in sight—the fry pits, the big-box stores, the office units, the lube joints, the carpet warehouses, the parking lagoons, the jive plastic townhouse clusters, the uproar of signs, the highway itself clogged with cars—as though the whole thing had been designed by some diabolical force bent on making human beings miserable. And naturally this experience can make us feel glum about the nature and future of civilization.[27]

When we think about the concept of the built environment and how it can promote our mental health, we need to consider how the design of our communities can effectively, safely, and inexpensively promote healthy choices. At the core of this idea is motivating change through planned environmental conditions. Those who work in public health know quite a bit about changing people's behavior. For example, they have instituted programs that have reduced smoking and increased the use of seat belts.

There is a set of techniques called *social marketing* that help to induce cultural change, making the change popular, convenient, easy, and fun. Communities and whole societies can bring about change by basing it on values and by framing the issues well. If we make a healthy

Figure 2.7 Communities can be segregated by pricing as well as location.
Source: Graphic by Scott Izen for the Media Policy Center.

the natural environment, so some spaces will need to be indoors in communities where there is snow for 80 percent of the year, and some communities will need bandstands in the park and shady trees to capture warm summer breezes. Our social health comes from the interchange that begins between the individual and others within the community and expands to a sense of the greater community.

We currently plan our communities to protect us from things we think we do not want to see (Figure 2.7). Our adult communities, for example, admit only people over fifty-five years of age. There are no schools so there are no young families. This monoculture is attractive to many who do not want to be bothered by the loud noises of a teenager next door or the nuisance of dogs barking from homes empty all day as families are at work and school. Yet there is another side to this picture. The couple who move in at fifty-five watch as their neighbors, and they themselves, become less active, then begin using canes, walkers, and wheelchairs. They have to begin checking on who in the neighborhood still drives a car so everyone can get groceries. The neighborhood sounds change from chatting across the porch to the rumble of the ambulance taking another neighbor to the hospital or the morgue. As residents rotate in a home, those who have been in the community longest may refrain from building close friendships. After all, the ambulance will be on its way again.

Social health goes beyond the quality of your neighborhood and the provision of services for the poor or underrepresented. When we segregate ourselves by where and how we live, we lose our understanding of

choice cool, then people are more likely to adopt it. In addition, through planning our communities we can make it easy for ourselves to do cool, healthy things by putting workplaces, recreational sites, restaurants, and stores near where we live. We can make exercise convenient by putting in attractive sidewalks that make us want to walk. As we create our built environment we need to adapt to each other. This vacuum of understanding is filled by assumptions. When people do not know or understand, they become fearful. What are the social health implications of violence and hate in the United States? The media capitalize on our inherent fears through their selection of news stories and the types of entertainment programs they create. Fear and hate are social health problems with tangible effects.

What does it mean to have a healthy society? What are the value judgments inherent in this concept? Building a healthy city requires a clear vision of healthy living and a narrative that can communicate that vision to others. This vision must address the cause-and-effect nature of our built environment and that environment's impact on the people who live, work, and play in it.

Chapter 3

Can the Built Environment Build Community?

In public health, epidemiologists study people and places to look for what is causing diseases and for ways to reduce and even eliminate them. They analyze data, breaking every question into answers by category, like age, sex, and geographical location. In their talks they show charts with data meticulously broken down. They "slice and dice" statistical analysis into dozens of categories in the search for the underlying causes of a particular disease, hoping to uncover opportunities for upstream prevention.

Yes, while epidemiologists, including myself, mine data and pursue mathematical reductionism, I assert that too often they take for granted the health importance of income. If you are poor in the United States, you are expected to have relatively poorer health. This is such a strongly internalized reality in our country that we are surprised when it is not true in Costa Rica or the state of Kerala in India.

There is another big influence on our health, and it is *place*. Blood pressure, obesity, physical activity, violence, and car crash risks—all these factors are rolled into the places where we live, learn, and work. Two of the most important new tools to improve our health are not conventional health tools at all; they are geographic information system (GIS) devices and cell phones. These tools tell us important information about where people spend their time.

When I was trying to do birth defect studies of farmworkers' exposure to pesticides, I had to rely on questionnaires. Things have changed a lot in California since then. Now all agricultural pesticide use must be tracked and reported to the California Department of Pesticide Regulation. Just as the farmer is using a computer and GIS to map out crops and water and chemical use, so too should we be able to track where farmworkers are working or being exposed. This is where the cell phone comes in. Everyone knows how easy it is to take a snapshot with most cell phones and to store or send it. So also, with certain applications, one can track in which fields and on what days the workers were present. The cell phone can become a tracking device, and the day is not far off when we will be able to identify actual chemical exposures by using electronic chips embedded in a cell phone. This technology is coming out of some of the terrorism preparedness and detection work.

For a farmworker, her health, and her baby's health, is affected by nutrition and medical care, but it is also very much affected by *where* she lives and works. This is true for all of us, even if not always as obviously. We are connected in a series of layers—layers of age, race, gender, occupation, and lifestyle. Although each one of us has the life goal of being more than the "sum of our parts," these parts do really matter in making us who we are, and our stack of parts tips us toward health or illness. The built environment is a critical layer that influences our occupation and lifestyle, those layers that we use to define ourselves.

ORGANIC PLACES ARE HEALTHY PLACES

The built environment is planted in a geographical place. It needs to belong to that place. When we go to cities built before technology made it easy and apparently cost effective to move large amounts of brick, concrete, steel, and wood over considerable distances, buildings and towns were almost always built with local materials, and the climate dictated the roof overhangs, thickness of the walls, water systems, and more.

The lovely size and shape and the functionality of the roofing tiles on California missions make the architecture special (Figure 3.1). These buildings usually nested into the local ecosystem, and if they were to survive, they had to disrupt the "nature" of the place (in both senses) very little. After a few years, they looked as though they had always been there. I think that architecture should be

Figure 3.1 Traditional roofing tiles on a California mission.
Source: Photograph by Stacy Sinclair.

right for the place and right for the people, and much less about the egos of a building's architect and its investors.

When my family and I visited Italy, we saw how on warm evenings everyone was out in the square to take advantage of the lovely weather, enjoying their *passeggiata*, their stroll, through the town. Old folks were talking, having dessert, and playing cards. Groups of women were chatting. Teenage boys walked around the square in one direction and teenage girls walked in the other, meeting and passing each other. There was music, people laughing, the smell of food and flowers, and this combination of the built environment and culture created an organic ecosystem that was productive and fruitful. Through choice of building materials and design elements,

Figure 3.2   Strolling in a public square in Italy.
Source: Photograph by Dick Jackson.

old communities more than reflect their sites—they are places of the heart that resonate harmony (Figure 3.2).

If the buildings and the cultural practices where we live are organic to that place, they fit. They are nurtured by the natural patterns, the weather, the seasons, and the life spans—they feel *right*. They are more durable and have less negative impact than something imported from the outside.

Layers of Communities Interact with the Built Environment

We are all members of many different communities as defined by our ethnicity, religion, nationality, gender, career choice, family, school, friends, hobbies, interests, level of income, and values and beliefs. Membership in some communities excludes us from others. Community happens when people connect with each other. Community may be based on shared legacy and heritage. Some communities confer rights, privileges, and responsibilities on their members. Membership is usually considered a source of pride and security but at times may also be a source of embarrassment or confinement.

We tend to help those whom we see as a part of one of our communities. In times of need, we offer a bed or an introduction that can lead to work and success for someone who is a part of our group. In religion, these conformities can be comforting, and provide answers as to what is expected in various life situations, but there are also costs to membership; some may find it confining and at odds with other needs.

To be outside a community that is important to us, feeling that we do not belong, affects our mental as well as physical health. Having a sense of belonging, of being a part of desirable communities, is a necessary component of a healthy life.

The built environment is not a community. Community is the software for the hardware of the built environment, and as anyone who has struggled with an overloaded, outdated, or failing computer knows, both "wares" need to be working or the system crashes. The hardware, the way we build our neighborhoods, urban centers, cities, and states can make it easier or harder to feel the sense of community within a geographical area. We choose some of our communities and others are imposed on us; in either case, the built environment can entice us to increase our participation or can create barriers that we must choose to overcome.

New York has some great examples of how the built environment either entices or creates barriers to community. Fred Kent, president of the Project for Public Spaces in New York, describes two contrasting examples:

Chelsea Market is all about street life, even though it goes through a building. People gather and shop, they talk and build community. The High Line is about art and design—it is separate from the street and there is no connection to any other buildings—there is very limited dimension to it. When people are up on the High Line the street below suffers. If the High Line and the street had been connected, it would have become a real destination. Right now it's just something that's fanciful, a limiting experience for New York.

If how we build our cities can influence how we spend our time, we need to examine those constructs that bring people together or keep them apart.

Our Neighborhoods

When I visit New York City, I always plan a long walk with my friend Father Ken Boller. Boller explains that "New York City is the biggest small town in America," and it is! When I was in medical school in San Francisco and new to the West Coast, I would be asked (no doubt on account of my charming accent), "Where are you from?" I would say, "The New York area." One day a Jersey (we never say "Joisey") friend said, "Don't ever say that. You're not from New York, you're from New Jersey. Be proud of that." I blushed and stammered. He was right. I never again said I was a New Yorker.

How large is a neighborhood? Some studies define a neighborhood by its square footage, and others by the number of people living in proximity. Often neighborhoods are defined by a common organizing structure—all the buildings that use the same post office, the boundaries for attending a high school and its feeder schools, or boundaries such as city limits. Neighborhoods are what cities are all about, and I think culture resides in them too.

When we design our neighborhoods, the built environment can encourage interaction or hinder it. In older communities, neighborhoods are built around squares or parks. Each square or park has a place to sit in the shade and maybe a monument to a historical figure; there may be a basketball hoop or a place to toss a baseball. Some are large enough to have a pond that doubles as a water reservoir. In some neighborhoods, including some in Detroit, Michigan, people are planting flowers and vegetables in these squares for people to share. The surrounding buildings are both residential and business oriented. There are walkways and people can look down from their homes to see what is happening on the street. Many eyes help make the streets and open areas safer. Older people watch the goings-on and act as informal security for the children. Children play outside while others jog, walk their dogs, draw, or read a book. And they interact.

Thoreau in *Walden* says, "I never found the companion that was so companionable as solitude." And we all need respite from the chaos and challenges of day-to-day life. We do not want crashers at our daughters' weddings, and we dread sitting next to loud, ill-mannered people at restaurants. We all need to be protected from cheats and bullies, and some people choose to live like hermits because that suits their basic personality and mysticism, but most of us need our solitude in balance with connection.

And indeed, later in *Walden*, Thoreau, in striking overstatement, writes, "I have travelled a good deal in Concord." Concord was a small town. Yet the glory of small places is that by knowing a hundred well-etched personalities, Thoreau was able to get an excellent measure of the human species. What would Thoreau think about today's gated communities? I suspect he would be astonished and saddened. I think what underlies so many of our strange decisions in the United States is raging insecurity. What are we hiding from? Visit a modern subdivision, gated or not, and often the streets are deadly quiet and lined with rows of garage-centric houses. People drive into their garage, lower the door, and enter the house from the garage. They exit by the same route, rarely stepping onto the sidewalk. To buy an ice cream on a hot day you have to get in the car and leave the neighborhood to reach the strip mall. Children may know each other because they meet at school, but when do people

of any age meet informally and strike up conversations that become lifelong friendships?

When I think about my childhood I think I grew up feral. By age seven I would go out at eight in the morning and stop home for lunch, then go out again until nightfall. One such early January day, when I was in first grade, some friends and I decided we would make a fort in the backyard out of discarded Christmas trees. We went through the Newark neighborhood and found our construction materials in front of nearly every house—a true mother lode. In the course of the day we filled the backyard with crackly, aromatic, very dry trees. My hands were covered with sap and I have no doubt we smelled sweetly of pine. My mother (a widow with a lot on her mind) did not visit the yard until dinnertime. I am sure my grandpa Jackson was thinking, "One match, and the kids and the house go up in flames." He commended me on my diligence but did not say much more. The next morning I awoke to find the trees gone, with only a small pile of pine needles on the front curb. Grandpa never said a word, so at the time I just figured it was magic—maybe Santa wanted them.

Most people in my generation and kind of location had similar experiences. We had a lot of autonomy—and we could find ourselves in real adventures and sometimes in real danger—like being stuck on an ice floe in the river. The built environment has not changed as much as our culture has—and our "toys" and the way our children play (Figure 3.3). The adventure of getting lost is blunted by GPS devices, and the pleasure

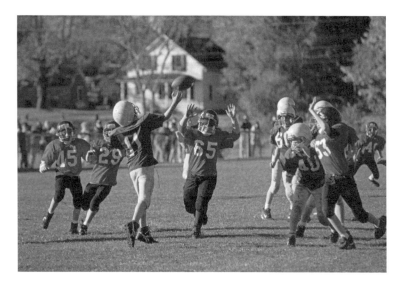

Figure 3.3 Children lost the opportunity to roam in only four generations.
Source: Photograph by Joseph Sohm. Used with permission.

of learning to read a map expunged by the dulcet voice in the dashboard calling out where to make the next turn. We went from being *free-range children* to being *helicopter parents*. And what is the impact on our children of having no real freedom?

Communities often build connectedness through the local school. What happens to the sense of community when children are bused across town or between towns? These hours in the car may provide opportunities for children to talk with their parents, but they are not opportunities for exercising, and the parents' time to prepare healthy meals in an already busy schedule is further reduced.

When I was young, I noticed that my mother's friends who were not family were the parents of my friends. Making friends is not just a convenience—it is essential. When my wife and I lived in Lafayette, California, our sons could bike to school. After their classes, they could stay there for activities and team sports. When they went to high school in Atlanta, the school was good, and ethnically diverse. There didn't seem to be much interaction between ethnic groups during the school day, but this changed after school; that was where friendships grew and a lot of the maturation of high school occurred. The problem was, most inner-city students had to get on the bus by 3 PM, so they could not participate in the band or most sports. We as a community deprived young, strong men and women of new friends, increased autonomy, and confidence because they had to get home by bus. Although busing was meeting one good need, that of diversity, the lack of neighborhood schools was depriving kids of lifelong friendships.

Through zoning and repurposing, we can create neighborhoods that reflect a set of intentional values. The Brewery in downtown Los Angeles is an example of an intentional community with shared values. A site with a number of old structures, including a former Pabst Blue Ribbon brewery and an old Edison power station, was redesigned as a mixed-use combination of artists' lofts, stores, and galleries. Catering to both traditional and avant-garde artistic endeavors, old buildings are repurposed to express the latest in modern architecture within small spaces, the use of sustainable building materials in action, and community interaction through intentional

School Architect Roland Wiley on the Importance of the Neighborhood School

Neighborhood schools are so important because they assure the sustainability of the community. The children's walking to and from school reinforces interaction with the neighbors, or it should. It creates that sense of community. When you have children getting in a car and driving miles to another school, the time it takes to drive the distance takes away from the life on the street, the pedestrian activity. The elders should be able to observe the young kids. The parents should know the other parents of children going to the same school. Community schools contribute to the health and well-being of the neighborhood.

design and landscaping. Those who choose to live and work there have built a community based on shared interests.

Why have we intentionally designed connectedness right out of places where people go to be together? When we go to many of the newer restaurants, there is no artwork on the walls. Instead we see multiple flat screen TVs with text (often phonetically spelled and incoherent) scrolling across the bottom. This is a big change in the built environment and a telling one. Certainly some marketing researchers have done a study showing that people drink more or spend more money with

more stimuli or these eateries would not be doing this. Personally (this does annoy my wife), I cannot hold a conversation with all that distraction. Meals should be important, almost sacred, times for conversation.

We are using the built environment to create barriers between us and we are building electronic toys and tools that isolate us as well. Think about a school day from an American child's perspective—eat breakfast in a box at home, get into a box to ride to school, sit in a box for classes, stand or sit in a box while eating at school, back in a box to get home, and watch a box at dinner or for recreation. We need a Box Rebellion.

Teachers have told me that children who do not practice having conversations have trouble getting along with others and doing their lessons in school. They are prone to aggressiveness and they get frustrated because they cannot find the words to express themselves. The cycle reinforces isolation and self-focus. We wonder why we see increasing aggression, depression, and attention deficit in our children. I think the truth is we have created it.

URBAN CENTERS

Urban centers have an opportunity to create a special kind of community, different from the community found in suburbs or exurbs. Rural areas seem to have both less and more community than urban settings. Less because one's neighbors are not living close by, and people don't routinely bump into each other while going about their daily activities. More because when it takes an effort to find people, no one can take others for granted. When those of us living in rural communities want to be with people, we seek them out—at the feed store, the coffee shop, the church, the school, or the band concert in the park. Urban centers, in contrast, are more the result of architectural design and city planning, and this built environment can enhance or inhibit the formation of community. (For example, pedestrian-friendly, street-side coffee shops, like the one in Plate 25, help people connect.)

In the post–World War I era, one of the most note-worthy architects defining the urban American landscape was Mies van der Rohe, a German architect. He and Harold Greenwald aimed to create a modernist style, akin to Bauhaus style, using industrial steel and plate glass to create simple structures that were efficient and focused on open space design. Dan Solomon, an architect in San Francisco, has postulated that this form of modernism grew out of our understanding of germ theory. We went from an ornate (some would say cluttered) decorative style, Victorian and Edwardian era rococo, to a clean, almost antiseptic environment—Bauhaus. Although the idea was at least in part to make a place that can be easily kept clean, the effect is austere and, for me, depressing.

In places like Lafayette Park, in Detroit, Michigan, these architects worked with landscape designer Alfred Caldwell to position houses facing each other and with green space for children to play while adults could watch or withdraw into their private space. The idea was to

provide a minimalist structure that allowed people to see both what was inside and outside the building. Glass was used to blur the line between the internal and external landscape. When we have a window overlooking a park, the trees are a part of our home and we are continuously engaged with nature. Hilary Robbie, who grew up in this area, says, "It's like living in a tree house on the second floor and a jewel box on the ground floor. You are always aware of the seasons, the animals, and the cycles of nature." This connection between our lives and nature is healthy.

Unfortunately, architectural ideas do not always have solely their intended impact. Even when they offer fine vistas of the outdoor world, these austere buildings are cold and often off-putting. In fact, some modern architecture is known as *brutalist*. Though that is certainly the feeling I get from many of the buildings from the 1950s and 1960s, this term is not intended to convey hidden brutality, but rather has its origin in the French term for "raw concrete," *béton brut*. There are a good number of beautiful buildings on the UCLA campus. Look at the UCLA Web site and brochures; all feature the lovely, red-brick Italianate Royce Hall, built in the 1920s, but nowhere will you find the cold, blank-walled boxes of the 1950s and '60s.

Watch children at play: they love to build forts and small, snug spaces. And as adults, we prefer to read and study in nooks and spaces with unusual shapes. We like rooms with tall ceilings. We don't mind a smaller floor space when there are higher ceilings. It does not cost much more to heat a room with twelve-foot ceilings than one with eight-foot ceilings. Our experience is that such rooms are more comfortable, though we are often unaware of why. Some people and groups in America, such as the Small House Society, are already advocating for and building smaller houses as a way of reducing environmental impact, gaining more enjoyable houses, and providing affordable and comfortable housing to more people.[1]

Sometimes when we are planning or designing, we forget who it is we are really creating buildings for. Our buildings are a legacy for future generations. Do we design buildings for middle-aged working people? Do we remember that young people, the elderly, and the poor have additional or different needs?

One of these days, if we live long enough, we will become disabled. They increasingly *got this* in Atlanta, the site of the first *visitability* law for the disabled. An important figure in this effort is Eleanor Smith, who is a leader in Concrete Change, an organization whose aim is to have "every new home visitable." The goal is that all homes, not just those designed for disabled persons, should be usable by and accessible to a person with a disability. Since the law was passed in 1989, about 600 houses have been designed that are fully accessible by the disabled. The uniqueness of this scenario is that these houses weren't originally built for the disabled. Eleanor Smith, for example, lives in a cohousing community that has sixty-seven of these homes. During the ten years since the community was formed, many residents have

experienced temporary mobility problems such as post-surgical impairments or have hosted older relatives, and their homes did not have to be modified to meet their changing life situations. For people with these homes, having a friend who is disabled doesn't require rethinking where to meet. He or she is welcome and has full access to the property. This should get us thinking about designing homes for universal access. How many condos and townhouses are built without a bathroom or bedroom on the first floor? What happens if we can't climb the stairs?

An examination of the homes we design can also uncover other biases. Michael Howard, a community organizer and executive director of a community teaching garden called Eden Place, was working with an architect on low-income housing for Chicago's West Side neighborhood, and as he looked at the drawings he couldn't believe what was missing. He asked the architect, "Where do poor people eat?" According to Howard, the rest of the conversation went like this: "He said, 'Well, where would you want them to eat?' I asked, 'Where do you eat?' He said, 'In a dining room.' I said, 'Well, why didn't you put a dining room in these houses?' It never occurred to him that the poor still need a place to eat. Imagine!"

Encouraging new economies can spark revitalization of a waning community, but it should never sacrifice character. Mayor Joe Riley in Charleston, South Carolina, revitalized his city's downtown but kept the gazebos and parks, the building facades, and the character of Charleston while modernizing the insides of the buildings and the infrastructure that allows people and cars to move through the city. (Plate 10 displays a portion of Charleston's remarkable seaside promenade and park.)

When I was young, I thought leaders were just the people who had their pictures in the paper and were on the news. Over my career, however, I have seen how organizations take on the personalities and character of the person in charge. Leaders really do matter. Once we also realize that everyone can lead, we can begin to assert responsibility and accountability for what is within our circle of influence. One really pernicious idea is that a person is a "born leader." Yes, strength, good looks, intelligence, and charisma help, but each of us can and must develop leadership skills.

Recently I was talking with a custodian at UCLA who was busy picking up cigarette butts. "Why do people do that?" he asked. "It's disrespectful. I'd like to have some kind of sign here." This gentleman has a degree of leadership and responsibility and was asserting his sense of accountability. Whether it is about making places better or countering the disrepair and ugliness we have been tolerating in our buildings and neighborhoods, we have to think about our cities. When we feel and act *disem-powered* or allow others to feel and act that way, we create bad places. We need to create places as good as the ones people want to visit.

Schools must be a built part of a city's revitalization plan. In Detroit, recently retired elementary school principal Joan Robbie talks about the school down the street. "Kids can walk to school and return to a friend's house until it's time to come home. They can do their homework together, and when they need help, they can share a tutor, an older child in the neighborhood or a college student from nearby Wayne State University. The tutors are learning how to teach and the children are improving academically. The school's ratings are on the rise and the kids are polite and conversational."

In a recent rating to identify the best American cities, one of the measures that brought a city to the top of the list was level of cultural liveliness. Interestingly, a high prevalence of small local restaurants was seen as an asset whereas an abundance of franchise eateries was not. In his books, author Richard Florida emphasizes the need for a vibrant cultural scene that attracts the most creative and productive workers to an area—especially young people. Lively, creative people—whatever their ethnic background or social preferences—add to cultural capital. We tend to dismiss nonfinancial capital, but social, intellectual, and cultural—and natural—capital are the seeds for financial capital.[2]

Over my career I've learned that *social capital* is the glue that holds communities together. People are healthier and live longer lives in communities with heightened social capital. After disasters these communities recover more quickly than others. Long commute times reduce the level of community engagement and social capital. In his book *Bowling Alone*, Robert Putnam asserts that every minute spent commuting means less time for connecting ourselves to our community.[3] The simple rule of thumb is that every ten minutes more of commuting equates to cutting all forms of social connection by 10 percent. More commuting means 10 percent fewer parties, 10 percent fewer meetings with others in the community, and 10 percent fewer dinners with family. Built environments can enhance or stifle social capital.

Whether it is the architecture or the economy, people are making connections based on a common idea or interest that keep them together. "What keeps me in Detroit?" muses social entrepreneur Toby Barlow. "I think it's the sense of possibility, the sense that anything good can happen. People focus on the bad stuff that happens, but Detroit is a neighborhood town that's got a strong sense of community. I like that spirit and that energy. I used to live in Brooklyn where they've already done so much. Here in Detroit, we're just getting started!"

STATE AND NATION

Effective Involvement

As we move from our family dwelling to homeowners association meeting, town hall meeting, city council session, and sessions of our state legislature and the U.S. Congress, issues of scale increase and we perceive that our influence is reduced. As with concentric circles though, each level is dependent on the others and affected by the work of others.

One technique for bringing a community together behind a program to cultivate a healthy built environment is the classic town hall meeting. Former mayor Ed Schock and current mayor Dave Kaptain of Elgin, Illinois, used this format to deliver information, receive feedback from the community, and cultivate social ownership from the people who live and work in the new, more sustainable community Elgin is building.

The impetus for Elgin's transformation came from community leaders, but the real work will be done by hundreds of involved citizens—by the community itself. The outpouring of people willing to give up their time for this effort demonstrates the old premise that oftentimes the citizens lead their leaders. I think in cases like Elgin's, many Americans across the country have been on the leading edge of the sustainability movement, and their local governments have been slow to catch up. I do believe governments must be cautious about embracing new ideas or strategies that are untested, but once some communities have demonstrated the ability to improve health by taking action, governments must drop their caution.

One key to successful public meetings is ensuring that they bring together all the interested parties, each bringing a unique perspective to the process. In other words, a broad view is a benefit and may keep public meetings from getting bogged down in details. Details will come later. Having a lot of experts in the room brings tremendous knowledge and interest to the endeavor. Creating sustainability is about many things, including jobs and economic development. Through this level of discussion our communities just might find the answers they need.

The further away from the neighborhood we get, the harder it is for an individual to represent all the viewpoints within his or her circle of responsibility. What we see come out of government is a compromise, and yet we are constantly surprised at the lack of focus proposed programs end up with in order to get enough buy-in to win the vote. At its best, government is a compromise of strong voices. What can government really do? Through the tools of tax incentives and regulations, legislatures and boards can encourage or discourage behavior.

Community values tend to be reflected in government policy—we vote for leaders who share our values. The built environment is the tangible expression of what was valued by the people who built it. Everything we look at in a built environment was a thought in someone's head at some point. We humans adapt so quickly that we soon start thinking that what we see has to be that way. But we need to remember that all buildings are temporary. To the degree that we can take ownership of our landscape, our place in its larger sense, we are happier. Some of us may think in terms of change once we have lived with the result of someone's brainchild, but others of us may still maintain a stake in that original investment, and thus it may take years to re-create places that are not working well for us. For many of us it is a process of trial and error.

After Hurricane Katrina the options were to rebuild New Orleans or to move all the residents. In the case of dwellings on a flood plain, not rebuilding might be wise, but that rarely is the choice. When we do rebuild, we almost never replace exactly what was there before. So at least we have the choice to change our policies based on the set of values that has emerged from our experience. We can use all we know about "smart planning," yet we also need to ask who the city will be built for.

Although we act locally, our nation is represented worldwide by our federal government and viewed in terms of its exports. People around the world judge the United States by the actions of its government. Only a few can separate us from our government and say, "We love Americans, but not the American government," or conversely, "We love America, but not Americans." The compromises (or absence thereof) that the government implements represent our national community. The current polarized atmosphere in the U.S. Congress accurately reflects the political divisions within the United States, and these contentious views are often encouraged by political pundits and the media machine. When we condone the actions and decisions of our elected officials, their decisions in turn affect the lives of each of us within the community.

My friend, Bobby Milstein, created the graphic in Figure 3.4 that shows how the purpose of public health is to make being healthy less of an effort, less difficult. When we communicate about public health issues, I believe we should distill them to their simplest form. We have got to get to the essence. This echoes the wisdom behind the old line, "If I had more time I would have written less."

When I was at the Centers for Disease Control and Prevention (CDC) I briefed Congresswoman Nancy Pelosi

Figure 3.4 The purpose of public health is to reduce the effort required from an individual to overcome health burdens.

Source: Graphic by Bobby Milstein. Used with permission.

Individual effort

Public health

on biomonitoring—the measurement of chemicals in people. She had instigated a large appropriation for a program, and she wanted to know how it was working. That was a fair request. So the CDC lab staff handed me a seven-page, deeply complicated packet of information for the briefing. I spent nearly a day (this is a lot of time when you have a problem list numbering in the hundreds) going through and trying to make it more concrete. I sent it back to the lab where they retechnicalized it. I walked into the briefing with a report that I didn't want to give. I asked to have the written report given to the congressional staff, and then offered my understanding in my best straightforward language.

The lab staff were annoyed with my simplifying; they felt I was being disrespectful to them. I was called for a while (not to my face), "Doctor Dumbdown." Yes, we had the technical information for anyone who wanted to see it, but the point is to communicate with people, not over or around them. It takes fifteen years or more to be a really

good *whatever*. When someone's talking from that level of experience, it is important not to be a dismissive listener. It's also important for the speaker to know his or her listeners and to use language that they will understand.

Let's look at two important areas in which local choices have a considerable effect on our health and the health of our communities.

Local Water Is a Global Issue

Clean, fresh water is essential to the survival not only of individuals but also of economies and civilizations. Wars have been fought over access to water, and if you compare the prices you pay for a twelve-ounce bottle of water and for a gallon of gas, you can do the math. Water is expensive, and getting more expensive all the time.

Water quality is profoundly affected by the built environment. Throughout history, human habitations were sited and survived in accord with the availability of clean water. Great cities flourished when fresh water was delivered in quantity and dirty water effectively removed, along with waste. Reservoirs, aqueducts, and sewer systems are essential to conurbation. These built structures have evident health benefits in terms of supplying potable water and also sufficient water for hygiene and agriculture.

In America about half of our drinking water comes from surface sources and the rest from groundwater wells. The great epidemiology story associated with the

London cholera outbreak in 1854, when John Snow traced the outbreak to the Broad Street pump, also exemplifies the dangers of pit toilets and ponded waste overlying groundwater sources of drinking water.

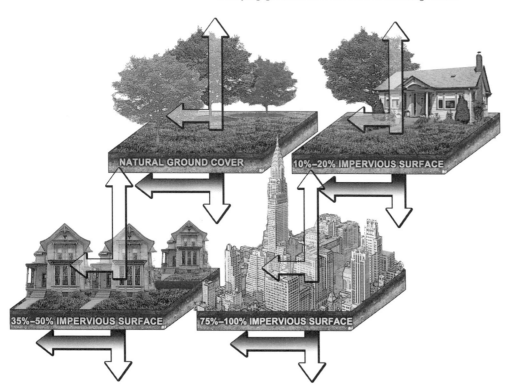

Figure 3.5 Runoff is directly proportional to surface material porosity.

Source: Adapted from "Relationship Between Impervious Cover and Surface Runoff" (fig. 3.21), in Federal Interagency Stream Restoration Working Group, *Stream Corridor Restoration: Principles, Processes, and Practices* (Washington, D.C.: U.S. Government Printing Office, 1998).

Moving water long distances is expensive in terms of human labor and use of fossil fuels. In general, water quality improves with increasing levels of percolation—note how bottled water is often marketed as having percolated for years through glacial fields and soils. In general, the more water that is captured where it falls, the better. The *base flow* of a stream is the water that flows long after a rainstorm ends. Water that has been allowed to penetrate into the earth seeps slowly out to water sources rather than running along the surface of the ground, removing topsoil and possibly causing landslides. A creek that is good for gathering that percolated water is good for people, or as a fisherman friend of mine said, every creek and river should be good for trout—if it is good for trout it is good for people.

As structures and paving cover the landscape, the amount of rainwater runoff increases. According to data from the Federal Interagency Stream Restoration Working Group, in a forest typically only 10 percent of rainwater runs off-site. In an urban area, about 55 percent runs off (Figure 3.5). Sometimes the health impact of changing the degree of water penetration has late-appearing and indirect effects. For example, paving over the cobblestones of the older streets in lower Manhattan led to the death of many street trees, and the loss of their shade and cooling.

In rural areas, well-managed, on-site water and waste systems (wells and septic systems) have minimal health impacts. When communities acquire more density and turn into sprawl, costly infrastructure to provide the water supply and septic removal is required. To reduce these costs many localities combine potable water supplies but require on-site treatment of waste by septic systems. These systems generally require homeowner diligence but often the only thorough inspection and maintenance occurs at the time a home is sold. In certain localities, septic system failure is reported at a rate of 5 to 40 percent. These failures lead to groundwater contamination.

Water also provides an important source of recreation. People need respite on hot days and often seek parks or coastal areas for play and relaxation. Unfortunately, in many parts of the United States, these places are among the most inaccessible, especially to the poor. New York City has 578 miles of waterfront but only a fraction of it is pedestrian accessible. (As Plate 9 illustrates, Charleston, South Carolina, is exceptional in providing access to its waterfront.)

Local Health Is a Global Issue

In this country we spend more than $2 trillion a year on health care. Ninety-six percent of it is focused on treating chronic disease. Research is showing us that perhaps 75 percent of all disease has to do with environmental conditions and broad lifestyle and social stressors. So, if we are spending 96 percent of our health care dollars on treating chronic disease, we are clearly missing an enormous

opportunity to start preventing these diseases. What we really need is a race for prevention.

We have got to go *upstream* and *prevent* some of the environmental exposures that are contributing to an epidemic of chronic disease in our society, and this is another area where the built environment meets community. We need an army of doctors and nurses, architects and engineers, public health and marketing people to focus their efforts on how we can build our communities to minimize environmental exposures and protect community members. To support wellness and a broader vision of health, new buildings need to model that kind of environmental responsibility, beginning with where they are sited and how they are designed, how they operate, how much energy they will use, and what kind of chemicals they will use for maintenance.

There are many examples in Europe of how to build and design hospitals that use half the typical hospital's current draw of energy. There are many examples around the world of small-scale, appropriate, preventive medicine. There is a lot that can be learned if we gather all these case studies from around the world and mine them to discern good ideas in prevention, methodologies for healing, and factors in environmentally responsible medicine.

We have increasing rates of asthma and obesity in America. Researchers are increasingly linking these diseases to environmental causes. Doctors are empowered and patients are better served when our medical teams are literate in environmental issues. When patients go to see their doctors, they should be asked about the cleaning chemicals being used in the home, school, and office that might be affecting air quality. The doctor should be aware of the local factories in the community and the roads on which diesel trucks might be traveling near patients' homes or places of business, affecting air quality in ways that might bring on asthma attacks.

We know that the more asthma attacks a person has, the more medications he or she needs, and the more his or her productivity at work or in school is reduced. When doctors know whether kids are driven to school or can walk there, how far a market with fresh produce is from the house, and whether there is a park or a garden where the family can get exercise and enjoy nature, a better diagnosis and course of treatment can be designed.

We know that exercise and daily habits influence obesity trends, and with an increase in obesity there is an increase in associated health problems, increasing medical care costs and, again, affecting attendance and productivity at work and school.

If we build walkable cities and protected bicycle paths, will we change the world? Maybe not, but those who live in these cities will live longer, healthier lives, and

they can better fulfill their potential. The solutions to our greatest challenges lie in that human potential.

At the beginning of the school year, I tell the incoming public health students at UCLA to go out of their way to get to know each other. They will learn as much from their fellow students as they do from their professors. Every week I call friends to discuss a theory or problem I am dealing with. There is no time like the present to start building this network of people who share your interests. Start early. It is a shame to discover someone you think is terrific a month before graduation.

I am going to give the advice that you knew was coming. We must begin our social engagement early. Be involved in local community issues. The world reflects the characters and ideas of those who have been involved in their communities, using the knowledge and technology at their disposal. If we do not like our community or if we think it is tolerable but could be better, it is up to us to create the community—and the world—that we want to live and raise our children in. Everything around us is the product of somebody's vision, somebody's ideas. Our community and the global community that it is a part of ought to reflect our ideas, not those of somebody who is long gone.

PART TWO
EXAMPLES OF CHANGE

Communities are ecosystems and they are always changing. As in any ecosystem, entropy happens. Systems will fall apart without a constant influx of energy. Sometimes positive change in a community can be managed by a set of strongly held beliefs or the tireless efforts of an exceptional leader. Some changes can occur relatively quickly, and others take systemic and gradual influence over time.

What follows are the stories of seven American communities, the choices they made, and the results that are emerging. A comprehensive list of communities that are changing across America could not be discussed in any depth here, but these examples depict some of the trends and choices that are symbolic of a movement toward health. It is likely you will see pieces of your community in the stories that follow, and I hope that they will plant ideas about directions that your community can take.

In Chapters Four and Five, you'll read about the Belmar district in Lakewood, Colorado, and about Prairie Crossing, Illinois. The builders and members of these communities had a formal intention to create a culture that reflects a set of beliefs. In the case of Belmar there was a community already in place, but circumstances created the opportunity for renewal.

When the economy changes so does the community, and the impact of that change can be tremendous blight. As discussed in Chapters Seven, Nine, and Ten, places like Elgin, Illinois, Oakland, California, and Detroit, Michigan, are looking at the impact of a shifting economy and are making choices to turn the negative into an opportunity for a newly revitalized community.

Towns with identity are destinations, and the vision and leadership of a few have created the distinctive lifestyles of Charleston, South Carolina, and Boulder, Colorado. As Chapters Six and Eight reveal, although these cities are very different communities that reflect very recognizable identities, a long-standing vision and persistent leadership along the path has made each unique for tourism and for those who live there.

What does it take to redefine a community's identity? In places like Elgin, community leaders are partnering with schools to raise young people who accept ownership for the community they took part in creating. In Charleston and Detroit, leaders are sifting through the historical importance of the community, maintaining the character and charm while retrofitting the infrastructure for modern conveniences and simultaneously sparking entrepreneurship.

Every community has a life story, and none of the stories of these communities has come to an end. By exploring where they are along the continuum of their adventure, you can see what is working and what may need further refinement, and what is more, I believe you will see yourself and neighborhoods like yours in the pages that follow.

Chapter 4

From Monoculture to Human Culture

The Belmar District of Lakewood, Colorado

Culture can nourish people in a place or leave them enfeebled with social and environmental malnutrition. Place, as represented by the built environment, and culture interact. When we think of Paris, Venice, or New York, for example, images come to mind. Maybe it is the look of the architectural styles reflected in the neighborhoods or the memory of historical events that took place. Perhaps we think about local foods, fashions, attitudes of the residents, or exports. Such characteristics of a place reflect and amplify its *culture*. We can often name some examples of things that go into a culture, but the sum of the parts, the totality of the details that give a place a unique feel or essence, may be difficult to define.

Culture is the set of behaviors, customs, and characteristics of a people. The term is also applied to the arts and institutions of a society. Culture often reflects what those who lived before us set up as important and valuable, and these values became policy and often law. Culture becomes embodied in what we build, and once

structures are built, they in turn shape those who inhabit them. This reverberation is reflected in how we use our public spaces, why people choose to live and work in some communities over others, and the businesses that enter and thrive in particular locations.

When I think of Belmar (Figure 4.1), the British idea of a *cultural industries quarter* comes to mind. Cultural industries quarters are intended to bring together a cluster of businesses that reflect a local identity, such as local artistic styles and materials. Belmar came out of an opportunity for renewal—a community decided its members were ready for something different and pulled its resources together to create a destination that reflects a new cultural identity based on vision and partnerships.

Let's examine Belmar as though it were a patient under the care of an attentive physician. When malls were being built in the 1970s, they were out on the edge of town, and they were happy to be out on the edge. Downtown was way over there and they were looking

Figure 4.1 One of the structures in the Belmar district of Lakewood, Colorado.

Source: Photograph from the Media Policy Center.

Lakewood, now called Belmar. By the time Belmar is completed, its twenty-two blocks will contain over a million square feet of offices, 175 retailers, and 1,300 condos, townhomes, and lofts. (One of Belmar's many mixed-use, downtown locations is displayed in Plate 4.)

SYMPTOMS

At the beginning of the twentieth century, wealthy people were moving out of downtown Denver and building estates in the rolling hills just outside the city. Frederick Gilmer Bonfils and May Bonfils, publishers of the *Denver Post*, owned one of these estates. The Bonfils mansion, sitting atop a few hundred acres, was known as Belmar. The history of this place is not unlike the history of other suburbs around the United States.

Following the Great Depression and World War II, higher-income families wanted to get away from the pollution, noise, and crowding of the cities, so the business of developing suburbs became increasingly popular. Over time these suburbs became less of a refuge for the wealthy and more of what we now call *suburbia*. Developers, eager to maximize land investments and achieve quick returns on their investment, built as many homes as possible on distant and comparatively inexpensive land, preferably served by highways. Building costs could be minimized by choosing inexpensive materials and using them in bulk, by reducing energy efficiency and

at prairie or whatever out at the other side. They were dependent on customers with automobiles, but gas was cheap; they loved their places. Then they were leapfrogged by housing developments and more malls, then they were leapfrogged again, and each new mall that came along took away market share, but ironically, this process of leapfrogging now often means that today a dead or dying mall actually has a central position in its metropolis. (Plate 24 displays an example of a dead mall, this one located in suburban Atlanta.) One such gigantic dead mall was Villa Italia in Lakewood, Colorado, now an inner-ring suburb of Denver. Today the site of the former mall is the home of the new downtown of sidewalk and park access, and by creating large schools designed to serve a wide geographical area.

Families moving into the suburbs wanted their children to be safe from traffic, and so they often opted for streets that were cul de sacs (Figure 4.2; also see Plate 22). The rest of the streets in these car-dominated neighborhoods were feeders and arteries intended to move many cars rapidly. So communities were planned with the maximum number of cul de sacs. Yet this apparently desirable design has created unintended isolation for children and the elderly.

After the war the country prospered, and Americans began building new businesses, attending to their

Figure 4.2 Bird's-eye view of a suburban neighborhood with cul de sacs.

Source: Photograph by David Shankbone. Used with permission.

families, and enjoying leisure time. Economic advancement allowed families to save for a home, but in choosing that home, they wanted something that reflected upward economic mobility. A home outside the city center reflected ideas of wealth, and the new planned communities promised quiet and calm surrounded by neighbors who shared their economic, ethnic, and religious beliefs. Urban areas emptied in part because of the lure of a better life in the suburbs, then eroded further due to declining tax bases, neglected infrastructure, racial shifts, and in some instances, the *coup de grâce* of an interstate highway system crisscrossing urban neighborhoods.

To build these enclaves required land, causing developers to move ever further from the city center to find it. Developers then created lakes and shopping centers to attract customers. With wealth came cars, and with cars came the freedom to travel greater distances. It appeared that communities no longer had to be close together and people no longer needed to live near where they worked.

Because of the distances involved in navigating the suburbs, getting the husband to work and the children to school, completing the chores, and maintaining the home became a logistical challenge. If the income could support two cars, the pace of life could again relax as more responsibilities could be split between the spouses. Soon the status of the *two-car family* took hold, and the type of car became an additional status symbol. It was better to have a Buick than a Chevy, and when you had really made it, a Cadillac. Sprawl is in many ways the result of our unspoken class system, and it has had unintended consequences.

Building Lakewood's First Shopping Mall

So the suburbs grew, and the Belmar estate became part of the newly incorporated city of Lakewood, Colorado. In the 1960s, real estate developer Gerri Von Frellick leased a portion of this estate and set about creating one of the largest indoor shopping malls of its time, Villa Italia. By 1969, Villa Italia was a part of Lakewood. It was famous for, perhaps ominously, "the largest parking lot in Colorado."

Though only four and a half miles from the businesses and stores of downtown Denver, the Belmar area grew into a wealthy community driven by the retail power of Villa Italia. During the 1970s and 1980s, this was a place people wanted to be, but as the families who lived there aged, new families did not come in. A community can be thought of as a living organism; its older cells must be replaced or it becomes frail. Without new families the neighborhood grew stagnant. Social and cultural malnutrition were becoming evident. Once the development was completed and the properties sold, the original developers had little or nothing left to do but move on to the next piece of open land and do it all again.

So it happened that by 1994, promised improvements to Villa Italia had not materialized, and cornerstone retail stores had consolidated and closed. Visiting a mall with closed storefronts became depressing, and Lakewood had to make a decision: What do we do with the property?

Not too far from Lakewood is the City of Englewood, Colorado, which also has gone through numerous regenerations going all the way back to the beginning of its history. Englewood was incorporated in 1903 and eventually it too would become infested with a mall. The land on which this mall was eventually to be built was home to an amusement park, an airplane factory, a film production center, and then Englewood's city park. In 1968, it became the site of the aptly named Cinderella City shopping center (analogous in the end to the progress in that fable from pumpkin to Cinderella's coach to pumpkin), built by the same developer who had built Villa Italia at Belmar. At that time, Cinderella City was the largest shopping center west of the Mississippi, with 1.3 million square feet, 260 stores, and all the parking that goes with that. Many people from all over the Rocky Mountain area—from Wyoming to Utah and Nevada—came to Colorado just to see Cinderella City.

The mall served Englewood by providing over 50 percent of the city's sales tax revenues. Some felt this success would continue forever, but as other malls got more exciting, revenues dwindled from their high back in 1968 to around 2 percent of city revenue in 1997.

By 1997, the mall had only one or two stores still operating and its infrastructure had begun to collapse. It became an eyesore. The city still wanted the excitement of having a special type of environment, so its leaders took a new route and opted to replace the mall with something similar to what's often called a *transit-oriented development* (TOD). And with that decision, Englewood began to develop a mixed-use type of development on fifty-five acres, using the key component of its new light rail station (see Plate 3) to reconnect the suburb with Denver.

"We used bond money to develop a piazza adjacent to the light rail station. But actually, we developed a whole new excitement for this area that has become the city center for Englewood," says Englewood city manager Gary Sears.

The success of the light rail system has been a factor in the excitement too. Many in the community initially thought the trains would bring *undesirables* out from downtown Denver, would attract graffiti, and would be dirty and underutilized, but this has not happened. As Gary Sears explains, "These trains were less like New York's subway and more like the systems in Washington, D.C., or Europe. People like having transportation options. They like being able to take the train to a sporting event or theater. Further south, they can hop in their cars and drive to other entertainment and environment destinations."

In creating a TOD a city needs to establish an environment that is supportive of the population, supportive of children coming and playing in the fountain, and supportive of residential life. People need to be able to walk to various restaurants and to get services, and the development must have a public presence. Englewood has had many representatives of other cities—Charlotte, Seattle, Austin, Minneapolis, and even Sydney, Australia—come to see what it has done.

It would be a mistake to think that developments such as Englewood's TOD or Lakewood's Belmar will go through without opposition. All redevelopment efforts meet with opposition from those who have other economic interests or those who want to thwart the success of a program that they oppose on ideological grounds.

Efforts to bring in public transit are always opposed by those who believe that more roads, more sprawl, and more of what we've had for the past sixty years are better than this alternative.

Thomas "Tom" Gougeon, Belmar's chief development officer, describes the transportation focus for the Belmar project this way:

We never actually labeled it a TOD project because it isn't on a main rail corridor. Our approach was to build a denser, more diverse, more walkable place where transit could work and be much more relevant and accessible. Even though numerous bus routes passed right by the 104-acre site when it was a mall, they all stopped on the perimeter of the site, on big, pedestrian-unfriendly arterials. People would then have to cross a huge expanse of surface parking to get to a destination. This was not an environment conducive to transit usage. We worked with the Regional Transportation District (RTD) to route the buses through the heart of the mixed-use district so residents hop on buses in the morning to leave Belmar for work elsewhere and employees at Belmar shops, restaurants, and businesses arrive and leave by bus. Teenagers use the bus to go to movies or ice skating on the plaza. Ultimately, we would also like to make a rail (probably trolley) connection down the right-of-way to connect Belmar to the West Corridor light rail and the entire regional rail system at the Federal Center stop. It would only require a streetcar connection of about 1.5 miles.

Recognizing a Dying Community

In pediatrics I remember being told to always listen to the older, more experienced nurses, and if one says, "Doc, I have a child you need to see *right now*," you go right away! Of course the nurse was always right. When I was an intern I asked a senior nurse, "How did you know that child had meningitis? It seemed you made the diagnosis from across the room." She told me, "If the child is limp in the mother's lap and not paying attention to what is going on, that child is *really* sick."

Maybe this is true with a community. If the people are listless and indifferent to their environment, that community is *really* sick and treatment is needed fast. We can look to nature to understand how our communities develop and change over time. Monocultures disintegrate in nature unless high inputs of resources and energy are continuously provided. Biologists examine ecological succession as the natural progression of an ecosystem. Initially, organisms move into an area and populate it because they are well suited to its present conditions. By harvesting nutrients and populating the area, in time they make the environment less fit for themselves but well suited for another set of organisms, who move in and further change the landscape. Rather than eliminating the "pioneer" species, the newcomers just spread over more and more of the landscape. In time a wide range of plants and animals share the space and a relatively balanced population emerges. As this community stabilizes, it thrives.

I would argue that all communities have an ecosystem as they have health and disease. As a physician I used to follow a set of procedures to take care of a patient. In public health the patient is the whole population. If our city has a disease, we need to find and treat the root causes, not just the blister or the fever we can observe. A healthy city needs interactive diversity—economic, ethnic, age, and social diversity.

Where there are young and artistic people with lots of energy, active days and a lively nightlife usually follow. In contrast, I have been told by residents of some of the gated, "golden age" communities that it is more routine for them to see a hearse than a baby carriage. Where there is a mix of young and old, that mix of generations makes the community a place to spend one's whole life, from cradle to coffin, and not just a place to park for a few years.

What happened in Lakewood's Villa Italia and in a number of other communities was failure due to monoculture. The construction of these suburbs rendered them inhospitable to populations that were unable to take advantage of the automobile-dependent lifestyle—those populations include children and the elderly, the disabled, and any family that could not afford a car for each driving-age person. So when *autoland* expands exponentially, what happens? Residents become isolated, immobile, and depressed. Imagine trying to navigate a typical 1960s suburb with broad lawns and no sidewalks if you are in a wheelchair. Imagine being an elderly person who must walk to the grocery store and carry food home without a car. It is a situation that is perfectly designed to be difficult and to alienate these populations from the community where they live.

When we live in an autocentric community with long distances between residential and commercial areas, what are the consequences to children? They must wait for parents or a friend to drive them, or they do not go far from home. When there is nearly nothing within walking distance to interest a young person and it is near-lethal to bicycle, he or she must relinquish autonomy—a capacity every creature must develop just as much as strength and endurance. And instead, when these cage-raised children do get out, they burn fossil fuel instead of their own body fat.

Where do these children go? A mall cannot be the downtown of a community, even though in many places the mall is all there is. Most malls are owned and operated by a management company. They are private property, with signs that limit "loitering" to keep kids from hanging out.

What about using public transportation? Often the bus stops are not close to where children live, and the places they want to go are not along the bus route, so long walks are required. The destination that can be reached is downtown, which parents often see as dangerous, a place they worked hard to get away from by living in the suburbs in the first place.

What about travel close to the modern American home? Without sidewalks, children have to be careful not to get hit by passing cars. They stay near the house and live their lives through the Internet. Jack Neff, in an article titled "Is Digital Revolution Driving Decline in U.S. Car

Culture?" points to the considerable decrease in licensed sixteen- and seventeen-year-old drivers. Where 50 to 75 percent (depending on age), of teens were once licensed, now the percentages are at 31 to 49 percent.[1] This is the first American generation since the automobile began to be manufactured that is less interested in driving than the previous generation. Why aren't young people waiting anxiously for their "freedom through wheels"?

In some ways this is a good trend because teens are neurologically unready for driving, and that is not due merely to a lack of experience. Beyond all that, what happens when the nearby mall and grocery store close down because newer and more fashionable malls and stores open up miles away? It does not affect the automobile user much, but it is devastating to those without cars. Indeed, drivers can live in a suburb of tens of thousands of people and still interact with only the same few humans year after year.

In Lakewood, according to Tom Gougeon:

There was a lot of nostalgia and affection for the old mall, but for the decade prior to redevelopment, it had been in decline and was each year becoming a less attractive, scarier place. We were very careful in how we approached Lakewood citizens. We first helped them understand why the mall had declined, how the retail world works, and why the mall wasn't going to come back to life as a mall. Then we engaged them in a discussion about what should take its place. We talked with them about the kind of environments they liked and didn't like. In the end, a lot of what they described were some of the better pedestrian environments in the Denver/Front Range Colorado area (Larimer Square in Denver, Boulder's Pearl Street Mall, downtown Fort Collins, and Lower Downtown Denver). We showed them the scale of buildings, dimensions of streets, amenities, and mix of uses that characterize those places. After about a year of discussion, we essentially had a cross-section of people endorsing the idea of a mixed use, walkable, transit-oriented district, really a downtown for a community that had never had a downtown.

In the end, I think, most people thought the idea for Belmar came from the community. There was a lot of local buy-in and really no opposition group ever formed. This is remarkable in a somewhat conservative community that is often fearful of change.

DIAGNOSIS

The mayor of Lakewood engaged his community in deciding what direction Villa Italia should take. He understood that involving people in the "work of citizenship," including meaningful discussion about their neighborhood vision, would provide the components that make a place interesting and livable. (Fred Kent, president of the Project for Public Spaces, has developed a protocol for helping communities decide what they want. He calls

it the *Power of Ten*.[2] If you ask people what the ten best and ten worst places in their neighborhood are or what their ten biggest opportunities are, they come up with amazing things. In a community, Kent says, "people want what is real, something authentic and alive, and they can recognize it right away.")

This redevelopment project became an opportunity for people to be a part of the history of the town. The citizens of Lakewood decided to start over. A variety of factors contributed to their desire for a different suburban pattern. The demographics of the community had changed and so had the values. The typical household today in the United States is very different from that of the 1960s. All the freedom and privacy that the typical suburban community provided was a trade-off with the time spent driving. Having to make so many car trips is no longer appealing, and people have gravitated back to environments that are much more like those that were popular before World War II.

After almost fifty years of Villa Italia, many citizens found it difficult to let go of the mall. Lakewood wanted to create a downtown for itself, a suburban city that never had one. With active support from the city and the help of the developer, Continuum Partners, an advisory group started to envision the future. They wanted a sense of community—a sense of place. They wanted a place that encouraged pedestrians and interaction, a place for people to shop, live, and work during the day and in the evenings. (Plate 2 displays the pedestrian-friendly Belmar Plaza.) Wy Livingstone, of Wystone's World Teas in Belmar, chose

Belmar for her business because Colorado "is very special, one of the healthiest states in the nation." According to rankings compiled by the United Health Foundation, Colorado was the eighth-healthiest state in 2009, as rated by such metrics as smoking, obesity, people lacking health insurance, and children in poverty.[3]

Donna Plutschuck owns and operates a business called Corporate Office Images Executive Office Suites, located in Belmar. It has office suites available for professionals to rent. She chose Belmar for both practical and idealistic reasons:

I supply all of the amenities, conference rooms, kitchen, and receptionist. I also have virtual office services. When I first interviewed Belmar, they asked if I was into recycling—if I would mind recycling—and I thought, this is wonderful. It saves me making trips. By that time, I had already purchased a Toyota Prius and I was into recycling. Belmar has a Leadership in Environmental Engineering and Design (LEED) certified building, and it has reduced my use of electricity. They tore up the concrete, pulverized it, and reused it, so they're working in a very environmental way, and they continue to do so.

My husband and I purchased a townhouse here in Belmar. It's very nice. I used to live in Manhattan and I could walk to most places I wanted to go to. It's not a big deal if my husband and I both get home from work and we say, "Let's go out to eat." "Let's go to the gym." We have far more leisure time. I also have far

more time to volunteer. Where we lived before, I had to get into the car to do everything, so because of commuting, it was hard to really engage or contribute to my community. Now, we have the energy and the time.

What emerges from looking at Belmar's design and talking to its residents is the concept of *quality density*. People think Americans do not like density, and we truly and rightly do not like *bad density*. But people spend lots of money to go to places like Paris because they want to be in a culturally lively, thriving place—and it does not hurt to have good food and café tables at the same time—and they don't mind when that place happens to have density.

June Williamson, associate professor of architecture at the City University of New York School of Architecture, has been studying Belmar. She observes that "when Belmar is completely built it will have three times the density in terms of built area than what the mall that was there before had, but it will also reflect a mix of retail, offices, and a host of residential choices. It will be a complete transformation from a retail, car-dependent, single-use facility to a mixed-use, walkable town center."

I am sure that local doctors around Belmar who advise, "don't smoke, eat smarter, exercise and walk more, make your bones bear weight, use your balance, be with people, and stay active," see many more patients following that advice now than local doctors saw in that same area thirty years ago.

In 2001, the Congress for the New Urbanism found that of 2,000 malls studied, approximately 28 percent were dead or dying. Although Belmar is still a bit of a "teenager," it has a lot in place: 800 households of people living there; seventy shops, restaurants, offices, and cultural facilities; and it's still less than two-thirds developed. It takes a long time to realize a project like this. Of course the current economic downturn hasn't helped.

CURE

To implement the vision for Belmar required big financial commitments. The City of Lakewood found grants,

Figure 4.3 Wide sidewalks, recessed parking, landscaping, single-lane traffic, and bicycle parking support Belmar's focus on the pedestrian.

Source: Photograph from the Media Policy Center.

passed bonds, implemented tax incentives to encourage green technologies, lobbied the government for subsidies, and created partnerships with the public utilities to implement upcoming regulations.

To create the new Belmar, every aspect of the community needed to be coordinated: the infrastructure of streets and open spaces; lighting, water and electricity conduits; buildings for homes, retail, and other businesses; community gathering events and strategies. Within each of these categories, decisions about who and what to attract to the community had to be made.

Belmar is owned and operated by Continuum Partners and its entities, but the streets, sidewalks, and parks are owned by the City of Lakewood and are protected by the city. Continuum was granted maintenance contracts of the streets, however, so it could provide better than typical service, such as removing snow rather than plowing it to the curb. Most malls and planned communities own the land the buildings are on or lease it, but the streets remain the responsibility of the city and may suffer from limitations in the city services available.

The developers considered the streets and sidewalks an important factor in creating a space that focuses on the pedestrian. They worked with the fire district to change standards on street dimensions. Narrow streets slow traffic, improving safety for pedestrians and allowing improved visibility across streets, and wider sidewalks promote greater pedestrian traffic, increasing the viability of retail (Figure 4.3). The overall effect is a closer relationship between the street, the sidewalk, and ground-level retail.

To encourage walking and to reduce reliance on cars, buses thread through Belmar's streets, and plenty of parking is provided at strategic locations around the community. Parking in structures is free, whereas street parking is metered for short-term parking management. Public transportation reconnects people with places. The elderly and young people who do not drive can still visit and shop, engaging with the community on their schedule, on their terms. The sidewalks in Belmar are wider than other sidewalks found around the area and include landscaping and building elements designed to be unfriendly to skateboarders, who used to overrun areas of the old mall.

The developers installed public art, and planted trees and landscaped with flowers to return the natural beauty of Colorado to this piece of land. They installed a fountain where kids can gather and play. As I have discussed, after World War II people in the United States created a new kind of community. It was an auto-driven monoculture, and most of it was suburban. Over time, we have paid less and less attention to people's need for walkable places, so our current move toward retrofitting suburbia comes from the fact that we created places that really aren't sustainable and also that these places are not as attractive to people as they once were twenty or thirty years ago.

Development is still a market-based business, but there has been a big change in the orientation of projects and, frankly, in what people want. Real estate development can be a jumble of specialists. According to Belmar's Tom Gougeon, "there are people who did housing,

people who did offices and those who did retail. Nobody put those pieces together. Belmar is an example of what can work when the professions of design and planning and of banking and finance work together to make walkable, livable places again."

For example, at Belmar, the developers wanted a space to attract people during the day and evening at all times of the year. Having an ice rink seemed to be a solution that would draw people together outdoors in the winter months. Unique to this design, the height of buildings to the south of the rink was adjusted so they would act as a shade to keep the ice from melting, rather than building an overhead canopy that might feel oppressive or too enclosed. In warmer weather the ice rink becomes an outdoor café. Both environments provide a meeting place, promoting a greater sense of community.

Belmar's new buildings are LEED-certified and use local and environmentally friendly materials. Natural lighting and solar panels are employed, maximizing the utility of roof space on parking lots and other buildings. The planners, builder, and city fathers of Belmar set out to attract local businesses, including those that would create a "hip" factor for the development. Through a visioning process, a mix of desired crafts and retail groups was identified, and Belmar reached out to these artists, chefs, and boutique owners to help them find their way to Belmar. It offered subsidized rent to some to give them the financial leeway to start up something new. For others with established businesses, attractive lease agreements encouraged them to move their businesses to Belmar from surrounding areas. The idea was to create blocks within Belmar with particular emphasis. These blocks would create pockets of entrepreneurship and attract people with focused and common interests.

While working to attract particular businesses that would create the desired culture, Belmar also designed each block to marginalize those who were not sought after. Strategic use of lighting, evidence of safety and services, and stairs or the lack thereof attracted particular groups to certain areas within Belmar and dissuaded other groups from frequenting certain areas. Through a subtle balance of architecture, landscaping, and surveillance, the culture of Belmar began to form.

For example, at the design studio Idaho Stew, Stuart and Nicky Alden work for a lot of nonprofits and also corporations, from small to big.

> We work with kids and adults doing screen-printing. It's lots of fun. We work here, but we don't live here. This block was developed as part of the master plan to have working artists in the community. The planners felt that having an arts presence would add to the soul of the community. Belmar made it really affordable for working artists. We call it our playground. We get the chance to get our fingers dirty, to work with kids and different businesses.

Belmar has made a long-term commitment to helping new businesses get started, rather than focusing on immediate revenue. It is not enough to say you have a vision. Belmar is investing in individuals who will create a connected urban community.

PREVENTION

To prevent Belmar from slipping back into the underutilized landscape that it used to be will take a constant infusion of energy and vision. Or to put it another way, now that the social and economic malnutrition has been cured, it is necessary to maintain a state of balanced social and economic vitality. Other towns are taking an interest in Belmar's new urbanism and Englewood's transit-oriented development as examples of the new small-town America, where people know their neighbors and contribute to the community so it develops over time, not into something different but into something deeper. The key to building a healthy community is bringing people together and ensuring their power in the political process. Neighbors come to the library together or meet friends at the coffee shop. They sit around outside in the piazza and enjoy themselves as their children and grandchildren play. They take walks, ride their bicycles, and take in the beauty that combines the buildings, landscapes, sculptures, and lively activity.

Belmar is not targeted to any one sector or demographic. The community is designed to support as many different types of housing and price points as possible. The retail program focuses on the core customer base needed to support more than one million square feet of retail when the community is fully built. Diverse, walkable, transit- and pedestrian-oriented places are inherently more attractive to many and fit the lifestyles of both younger and older people. Belmar also works for singles and younger couples who want an urban lifestyle but may not be able to afford or want downtown Denver living or may have family or job reasons to be on the west side of town in or near Lakewood.

I commented earlier that it is the nature of ecosystems to find stability and that maintaining an unnatural state requires great influxes of energy. The built environment may seem somewhat static compared to the natural world, which is endlessly dynamic. Yet in an organic system, both are interacting intensively, and that is what we are really talking about in creating a new community— creating vitality. How do you create an organic vitality? By bringing the physical structures, the hardware, together with the metabolism of life. All buildings are temporary buildings. There is no long-term stability in

built environments. In nature, metabolism is the combination of anabolism (building up) and catabolism (tearing down). A healthy system has a balance of the two.

The city of Lakewood went through a cycle of catabolism and then anabolism in the Belmar area, and the community that is there now will eventually go through its own dissolution. Nature does not put something in place and then walk away from it. It will continue to replicate and evolve. Belmar is a lovely place, and its spaces to live, work, and play are not finished. Belmar will continue to build up and tear down at the same time, and community members cannot fight with that. They have to nurture it the way they nurture the changes their own children go through.

Now that Belmar is establishing itself, groups are coming from across the country and around the world to look at this experiment. Can Belmar be replicated? Will Englewood's transit-oriented development strategy work for other communities? People explore, study, and take ideas back to their communities and build their own visions.

Chapter 5

Using New Urbanism Principles to Build Community

Prairie Crossing, Illinois

Prairie Crossing is in Grayslake, Illinois, a suburb of Chicago.[1] Like the people in Elgin, Illinois (a community described in Chapter Eight), the people of Prairie Crossing are concerned with making a traditional, Midwestern American town sustainable. But where Elgin is retrofitting for sustainability, the people of Prairie Crossing have created a new kind of community that focuses on preserving land and building community through healthy living choices. The Prairie Crossing site is 677 acres of farmland within the 2,500-acre Liberty Prairie Reserve, so it is part of a permanent parks setaside. In the mid-1970s, the acreage was slated for typical development. In 1987, Dorothy Donnelley bought the property and with her nephew, George Ranney, and his wife, Vicky Ranney, set out to create a development with the same financial returns as a conventional development but based on a very different set of values.

In the early 1990s, Dorothy Donnelly and George and Vicky Ranney formed a group, Prairie Holdings Corporation, which became the overseer of the development. These three Prairie Crossing co-founders and developers had recognized that traditional developments were not providing the long-term economic vitality that conventional suburbs promised. As Andres Duany, Elizabeth Plater-Zyberk, and Jeff Speck argue in *Suburban Nation: The Rise of Sprawl and the Decline of the American Dream*, "the American Dream just doesn't seem to be coming true anymore."[2] Although suburbs provide conveniences, and often lower costs, they are burdened with a repetitive landscape, tedious commutes, and a growing sense that they have failed to provide a better quality of life.

The three founders turned to the idea of New Urbanism for a different answer. New Urbanism is based on principles that made some of our post-war conventional communities satisfying for residents. Communities based on New Urbanism offer *walkability*. Services are available close to residences, and narrow streets and wide sidewalks encourage walking. To emphasize this idea, garages are accessed from alleys. In place of a design that segregates services, residences, and businesses, multilevel and mixed-use streets are platted out with various building types, sizes, and prices in a mix intended to encourage economic diversity. Houses are close to each other and close to the street. In sum, issues of density, variation, harmony with the natural environment, and opportunities to interact with others surface when we examine the difference between traditional suburban sprawl and new urbanist designs (Figures 5.1 and 5.2).

Prairie Crossing evolved as a development of 359 sustainably designed, single-family homes, and 36 condominiums, built over one square mile of prairie. There are common buildings, a school, stores and restaurants, a train station for public transit to Chicago and O'Hare Airport, an organic farm, horse pastures and stable, and a market. The development site still includes more than 165 acres of restored prairie, 20 acres of restored wetland, and 16 acres of historic hedgerows. The community's Web site reports that "over 60 percent of Prairie Crossing is protected open land," and that "the community was designed to combine responsible development, the preservation of open land, and easy commuting by rail."[3]

Figure 5.1 Suburban sprawl.
Source: Photograph by Joseph Sohm. Used with permission.

Figure 5.2 A New Urbanism community.
Source: Photograph from the Media Policy Center.

Prairie Crossing was created to be sustainable, to "work" for the people who live in the community but also to work for the planet—to use fewer resources and to allow people to connect with each other. A portion of the residents work on the organic farms, grow their own food, and get to destinations easily using public transit. Could this be a pilot for future developments and subdivisions in the United States? The developers created a set of principles that they used to guide their decision making. Their principles are similar to those of New Urbanism and can be summarized as follows:[4]

- *Environmental protection and enhancement*. Construct greenways and place houses to protect the environment, native vegetation, and wildlife.
- *A healthier lifestyle*. Include opportunities for outdoor exercise in the design of the community; include a source of locally grown foods; provide local health care near where people live.
- *A sense of place*. Let architecture and the color palette for buildings reflect the natural surroundings and the history of the area.
- *A sense of community*. Bring together the homeowners, conservancy, public officials, and local businesses to collaborate on amenities to build, designs to approve, stewardship activities and community-based synergies to create.
- *Economic and racial diversity*. Promote a mix of incomes and ethnicities through cost structuring.
- *Convenient and efficient transportation*. Provide rail service to Chicago and O'Hare Airport; maintain major roads, routes to local schools, stores, and restaurants.
- *Energy conservation*. Construct homes to be 50 percent more energy efficient than other new homes in the area; engage residents in recycling and composting programs; encourage walking and bicycling as alternatives to short commutes by car; install a wind turbine to run the farm; build the local school to LEED standards.
- *Lifelong learning and education*. Support charter school programs focused on environmental themes; provide informal and formal learning settings for community members in varying age ranges.
- *Aesthetic design and high-quality construction*. Use the best choices available in design and building materials.
- *Economic viability*. Focus on long-term economic success for families and local businesses.

If we must build in green fields and we want detached housing, then Prairie Crossing seems to be the kind of community we need to be building. It is an example of a new type of community, where farm and city merge to bring the best of both worlds to people.

SYMPTOMS

Everyone wants to live in a safe place that is clean and attractive, but we also need to be around people we care about. We want what we do to matter. We need to rebuild the United States in a way that will work not just for the economy but also for people. This work needs to happen for middle-class families like those in Prairie Crossing and especially for the families in communities that are stressed financially and in the greatest need of support.

Prairie Crossing co-founder Vicky Ranney did not want to create a gated community, saying explicitly: "We wanted to create a place that was part of and embedded in the larger community. We are not about being elite—these are real people with real work."

Americans are struggling with obesity and trying to find healthy, fresh, locally grown food. I assert that we are in a perfect storm of social, health, environmental, and economic challenges. We have to think about solutions that cut across multiple domains at the same time. We know how to create places that make us healthy, that give us meaning in our lives, and fill us with vitality. We have got to go back to doing that. In Prairie Crossing, creating a community with a working farm began as part of marketing the development and was funded out of that budget, but the farm has become self-sustaining. A good organic farmer who does direct-to-market crops can generate $20,000 an acre. A more typical return for organic products, when there is no direct-to-consumer marketing, is $7,000 an acre. It is the difference between retail and wholesale marketing. It is not just fashionable but also good economics to go organic, especially if the consumers are close by.

If we are going to spend money, we ought to be spending it on things that work for the planet, our health, and our economy. We ought to build places that are places of the heart.

One of the big things affecting the planning of a community is its *general plan*. It is the constitution for the growth of the community. This plan determines where the highways are going to be, how many parks are needed, and how big the schools will be. (Plate 6 displays Prairie Crossing's development plan.) If this constitution describes how a community, neighborhood, or town is going to grow or stay the same, the planners will follow that plan. So, to change how we build communities, the general plan has to change. Making that change requires intervening upstream in the political process. Intervening upstream requires educating the planning commission so smarter developments aren't caught in local red tape.

DIAGNOSIS

We know how to create communities that are vital and lively because that is what we built a hundred years ago. Older communities required that people walk. Streets were narrow and there was little space to park a car. Somehow we have forgotten about the importance of vibrant town squares to the way people want to lead their lives—the idea that people want to have cafés and restaurants, farmers' markets, and little food stores where they can meet their neighbors and talk to the greengrocer. We have forgotten that people want to be in a place that has meaning to them, that reflects their personal history—where their children grew up and where their parents died.

When we think about the importance of *place*, the key is to identify what will make a place unique and memorable. Sometimes it is what has been left untouched, which can be as important as what has been developed. At Prairie Crossing the open land was returned to the original native prairie and wetland plant species that had covered the area prior to nineteenth-century cultivation to raise corn and soybeans. Removal of nonnative species allowed the reintegration of plants into a natural stormwater collection system, using long-rooted native plants to slow the percolation and purification of rainwater and snowmelt. Although the visual impact of the community is based on the landscape, the foundation from which this landscape emerged was creating a space for community—for connections between people.

Through a *learning farm*, or *incubator*, and a working farm on the property, people of all ages have been learning to raise crops and market their produce, sharing equipment and information. The produce from these farms is sold locally and is used by schools, stores, and market stands. As an integral part of the community, the farms foster interaction and opportunities for the community to work together to plant and harvest. Physical proximity and activity lead to healthy well-being.

We all need exercise to be healthy. In a large study of 100,000 nurses conducted over forty years, the nurses who stayed fit had a 60 percent lower death rate than those who did not, irrespective of weight gain.[5] Although some people like to go to gyms or fitness studios, many of us just love to walk, and it is a perfect exercise, especially as we age. No drug works as well as exercise for preventing type 2 diabetes from developing, but we need to walk a fair amount, about 10,000 steps a day, which is about five miles for me. It sounds like a lot, but spread across the day it is pretty easy to accomplish. More important, there is no drug as safe to help prevent diabetes—no drug as free of side effects to save your legs from gangrene or your heart from failing—as the exercise of walking.

Children need to walk or bike to school. Kids who walk or bike to school concentrate better. They behave better.[6] They're much easier to manage than kids who have been sitting in a box being brought back and forth to school. Every school ought to have a set of bike racks, and those racks should be full. Every school principal should encourage kids to walk or bike, and that principal should be supported by the parent-teacher association for his or her efforts.

The most prevalent disorder in the United States is depression. One effective treatment for mild depression is being with other people. For a hundred thousand years, human beings have gotten each other through hard times and dark winters, times of deaths and disease, by telling stories, being with each other, and having social connections. The other cure for depression is exercise—walking. Researchers have found that those who exercise in daylight and sunlight with appealing natural features around them will improve their mental well-being.[7]

When we look at our schools, we see a microcosm of the priorities we set in the greater community. In the early 1900s, we built schools that were iconic, that were a measure of how much we valued education. They were places that we admired, and they had a prominent place in the community. Beginning in the 1950s, we built bigger schools, and they looked increasingly like prisons. They were often located in a remote part of town because the land was cheaper there. Two-thirds of kids walked to school a generation ago. Now it's one in six.[8] Clearly, we have to rethink where we put our schools in relation to the rest of the community.

We knew how to build good classrooms a hundred years ago. In a good classroom the windows opened to get fresh air. The ceilings were high for better ventilation. There was natural light. The brain actually functions better in natural light than it does under fluorescent or incandescent light.[9] Kids learn better in these naturally ventilated, naturally lit rooms than they do under artificial lighting. We need to bring the outdoors inside much more. Taking the natural environment away from our students is unhealthy. There is a hum in fluorescent lights that young people can hear, and for many this hum can be distracting. We know this, and yet we have been building schools (and homes and businesses) that do not reflect this information.

Parents who commute far distances to and from work have little time to prepare healthy meals. Children grow up on highly processed foods that contain excessive fats, sugars, and salts. They do not know how to prepare healthy meals themselves, so these eating habits become an unbroken cycle. Schools have an opportunity to teach children new habits—habits they can take home to their families. If every school in the United States had a school garden, every child would learn how to grow fruits and vegetables. They could learn ways to prepare these harvests and try them as part of their school lunches. Children could get their parents to grow some food at home in containers or in small plots in the yard. Pretty soon we would see changes in what families eat. (Plates 5 and 7 show students engaged in a work-study activity at the Prairie Crossing learning farm.)

Children need to know where food comes from, and how good fresh vegetables and fruit taste. It is good exercise to plant, maintain, and harvest crops. Children learn about ecosystems, food chains, and photosynthesis. They can put the principles they are learning in the classroom into practice. It is fun to do, and schools can engage adults in the community to help.

In addition, schools are a natural community hub, a safe place, and a gathering place where people connect and build friendships. They are a safe place to learn about the community and the industries that sustain it.

The U.S. government subsidizes farmers to grow such things as grains (mostly corn) and cotton, but no subsidies go to the growing of fruits or vegetables. Many of the subsidies are not even for food crops but for crops used to manufacture products (for example, tobacco, wool, and cotton).[10] Rather than subsidizing our farms to *not* plant specific crops or to plant corn to make everything from ethanol to high fructose corn syrup, we ought to be subsidizing food that is healthy. At Prairie Crossing the developers contribute $10,000 per year from home sales to the farm program (the learning farm, the Farm Business Development Center, and Sandhill Organics) in recognition of the services provided to the community.

So, how do we build healthy communities? We start with healthy kids. How do we build healthy kids? We get them to exercise, to eat healthy foods, and to connect with each other. When kids connect with others through leadership, they can make the goals of healthy eating and exercising happen. Leadership is a skill people grow like every other skill. Our world is the way it is because someone in the past had a vision to make it this way. The world will be what it is going to be because we have a vision to make it the way we think it ought to be.

Children should begin social engagement early. They should be involved in activities and community leadership issues. They should learn to be involved in student government, making decisions about how their communities should operate and working to create the world that reflects what they feel is important. Children and teens are genuinely interested in learning how to grow food. There is health at the core of this activity. It is in childhood that we learn how to give back, and understanding how we can use the land to create food is a lifelong lesson.

Everything around us is the product of somebody's vision, somebody's idea—and our communities ought to reflect our ideas, not those of someone who is long gone.

CURE

The residents at Prairie Crossing know what their community is about. Vicky Ranney describes Prairie Crossing as "a conservation community, but it's also an economic investment. When a group of neighbors bought this land we were faced with how we would pay back the purchase price. We knew we had to develop it, but we had to figure out how to do it better than the people who were proposing to do it in a standard subdivision way. That was the beginning."

"We knew from the beginning we wanted to build a community that had a healthy lifestyle that improves the environment rather than damages it," said George Ranney.

We wanted to market the place based on how people chose to live. We're not professional developers ourselves, but Vicky and I have led these efforts over the years. We put together a team of people who are experts in development and finance, but also in the environment and how to think about land preservation and education. Because we knew we wanted lifelong learning to be one of our most important principles, we've kept most of those people together over the years. They are the team that carries us forward.

Land Preservation

One of the things that makes this development unique is the way the land has been treated. As Vicky Ranney describes it:

> We started with a landscape architect as a master planner. We worked from the very beginning with an ecologist who set up a stormwater system to create such clean water in our lake that people can swim in it and use it not only as a stormwater system but also as a recreational place for boating, swimming, and fishing. I think that connection between the people and the land, the health of the land with the health of the people, is very important.
>
> We wanted people to have a sense of place—a rootedness, a community that was built on the history of this specific place and the people who had contributed to the area. We felt strongly that the architecture and the natural history should work together. That's why we took this site, which had been a corn and beans farm, and replanted it to include the native prairie and wetland plants that had been here before the settlers ploughed it up.

On a patch of land that the Prairie Crossing developers could do anything with, they chose to save open land for natural areas and farmland. They chose to design the built environment with nature as a primary resource, for people to observe and interact with.

Prairie Crossing Housing Design

"We used to have a sign out front which said 'Built for the future, roots in the past,' and we looked at what had been done here in terms of architecture in this northern Illinois area," Vicky Ranney says. "A number of the residents came from New England and northern New York via the Erie Canal in the 1840s and that's why some people think Prairie Crossing looks like a New England village. We asked our architects to build in the local tradition, but where to build is something we decided."

Putting houses and businesses in a compact pattern makes the community feels neighborly. People sit on their front porches across the street from each other. The front yards are short, so people sit close to the sidewalk. The sidewalk is a place of life—where friends walk by. If you are on your porch you can talk to people walking down the sidewalk. You can see the children playing. It feels safe. It is an opportunity that builds community (Figure 5.3).

The streets are narrower than in a conventional subdivision, and cars parallel park along the streets. With narrow streets, cars drive slowly. After driving on the Interstate to and from work, coming back to this community is like arriving at a haven—a step back in time to a slower way of life. People can be close to the shops and train station, which are only two blocks away. This building pattern is similar to the way towns used to be built. The new-old design reemerges.

Figure 5.3 Homes at Prairie Crossing reflect both residents' new interests and their ancestral architecture.

Source: Photograph from the Media Policy Center.

The Prairie Crossing Way of Life

We know some things about the people who live in Prairie Crossing. Compared to the people who live in the surrounding areas, they are twice as likely to vote. Thirty-five percent of the residents take the train to work, 5 percent being the norm in the surrounding area. Additionally, they walk to the train. "There's a pervasive attitude about choosing what's healthy, physical, and intellectually demanding," says George Ranney.

The built environment affects the health of every family, and there are strategies that intelligent town planning can employ to support a natural way of living. The aim in Prairie Crossing is to make the built environment a supporter of an outdoor, healthy lifestyle. Ten miles of trails were built around the community. A lake was created as a place of respite, rather than as a drinking water reservoir, and it is great for recreation. People swim in the lake and bicycle along the trails. The community also has common spaces for people to gather. The indoor fitness center and community barn are used for a variety of activities.

Homes are energy efficient and designed for a natural flow. "What we've tried to do is put things together so they reinforce one another. What we end up with is a much more activist, healthy lifestyle than the norm in this general area," says George.

The Prairie Crossing Farm

Prairie Crossing is in the nation's bread basket. The conventional agriculture of corn and beans in the area is now a commodity crop, not a food crop. Residents recognize that when they go to the supermarket and get things from all over the world, the international trade is a luxury. People need to start growing their own foods locally.

The charter school supports the focus on personal independence and creating sustainable crops. Its curriculum is focused on getting children working in the natural environment and on the learning farm (Plates 5 and 7), learning their academics in a hands-on way.

George Ranney remembers the early planning stages:

We started with a farm here because Vicky said this would be a key to the community, and she was absolutely right. It's not only a signature element of the quality of life and an example of the healthy lifestyle we're promoting; it's getting people out into the fields and into the chicken house. I was skeptical of the farm in the beginning, but it's become something that brings families together, things for them to do with their children, a way to get to know the neighbors and to get to know the source of their food. The learning farm has become something that I think is not only a critical element of Prairie Crossing but the beginning of a national movement. Local food is going to be a way to get to a healthy lifestyle and we focused on it early. Our farm has become a motivating force in this eight-million-person region for how you can best feed people and involve them in the production of their food. On that score alone, our project has become more influential than we ever dreamed it would be.

We expanded the farm to include the school through a program called Farm to Table in which kids get to essentially take food that they are growing and bring it back to the table after it is prepared. Then together, we eat the food that students grew. This lunch takes place once a month and we feed about 200 to 250 meals. The Farm to Table program is significant because it gets at the holistic approach of the community, making the connection between the land, food, and people, and local economies and the health of the entire community.

Prairie Crossing eventually leased forty acres of land to Matt and Peg Shaeffer, who opened Sandhill Organics. The Shaeffers attribute their organic farming success to the local support system and the integration of their farming with the residential zones.

Does Prairie Crossing Make Financial Sense?

Prairie Crossing proves that people will pay more for quality of life, not just quantity. Residents want the trails and the community center and better quality homes, and judging by the sales prices of the homes, people are willing to pay 30 percent more per square foot to get it. Developments around the country are starting to understand that people want more than a roof. They want a way of life, and they are willing to pay for it. Some other communities that are operating on this premise and combining farms and homes in neighborhoods are Serenbe in Chattahoochee Hills, Georgia, just outside Atlanta; South Village in South Burlington, Vermont; Bundoran Farm in Charlottesville, Virginia; and Hidden Springs in Boise, Idaho.

"We are showing that people will purchase what they value. This is not a philanthropic project. It's a business venture. At first we were concerned that people wouldn't understand—that they'd think this was some sort of hippie fanatical community, but it's not. We positioned the community in the mainstream," said George Ranney. "People have bought here because it makes sense commercially. If you look around the country there are more than a dozen other communities being developed along similar ideas because they think it's good business." He envisions noteworthy social outcomes that will draw the interest of other builders and architects, even though he understands that long-term data on the health of the residents is not yet conclusive.

The housing downturn has been very tough for Vicky and George Ranney and for many other homeowners. Some folks in Prairie Crossing have lost their homes, just as in many other communities across America where residents are struggling financially. However, the big shift over the past two years has been in consciousness about food stemming from a concern for health and sustainability. Growing local food builds local pride and brings people together. The community among people living in proximity supports those who are struggling financially and together they are making it through.

PREVENTION

Prairie Crossing reflects many of the ideas that go into making a healthy built environment—it shows what the dedication and love of people for themselves, their

families, and their neighbors can create. It shows a commitment to healthy living practices, and the desire to connect with neighbors, to build community. Every home in this community uses *less than half* the amount of power, energy, and oil that a home built anywhere else does. Every home is near public transportation. The children can walk and bike back and forth to school. There is a local farm that produces organic food, and community members consume some of this food together at parties and events. Residents can have this food delivered to their homes as well.

Prairie Crossing is an attempt to connect people back to the land and back to a community—an effort to make the place where people live work for them, their health, and their grandchildren.

When President Obama addresses social initiatives, he talks about motivating, engaging, solving problems, and taking the solutions to scale. Prairie Crossing has engaged its designers and residents in solving some of their healthy living challenges, but can this solution be replicated?

According to George Ranney, "Yes, it clearly can. We've said to ourselves and other people that we would like to have other developers, other potential home buyers, look at us and say, 'Hey, we could be building a community like this'; so this is a model of best practices—an exemplary development from the beginning."

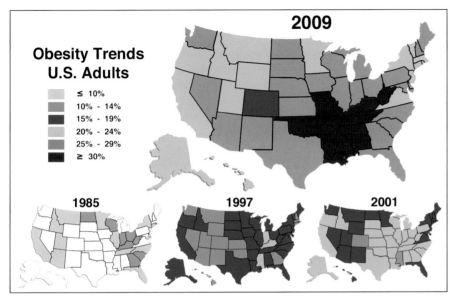

Obesity Trends U.S. Adults

- ≤ 10%
- 10% - 14%
- 15% - 19%
- 20% - 24%
- 25% - 29%
- ≥ 30%

2009

1985

1997

2001

◀ **Plate 1   The United States**
Since 1985, the fastest moving and most pervasive chronic disease infiltrating America is obesity. From 15 to 30 percent of the people in each of our states suffer from overweight, and our nation as a whole pays the health care cost burdens.

Source: Maps from Centers for Disease Control and Prevention.

▶ **Plate 2   The Belmar district in Lakewood, CO**
Healthy communities create places like Belmar Plaza, where people meet, relax, and enjoy the beauty that surrounds them.

Source: Photograph from Downtown Belmar Apartments, www.downtownbelmar apartments.com. Used with permission.

◀ **Plate 3   Englewood, CO**
Connecting Englewood with the established communities of Denver to the north and Littleton to the south, light rail gives people a public transportation choice and reduces traffic on the main north-south roads.

Source: Photograph from Media Policy Center.

▼ **Plate 4   The Belmar district in Lakewood, CO**
When they live in proximity to where they work and play, people interact more, building community.

Source: Photograph from Downtown Belmar Apartments, www.downtownbelmarapartments.com. Used with permission.

▲ **Plate 5   Prairie Crossing, IL**
When people work together and learn from each other, friendships as well as food grow.

Source: Photograph from George and Vicky Ranney. Reproduced courtesy of Prairie Crossing.

▶ Plate 6   Prairie Crossing, IL

Through specific choices, Prairie Crossing was designed to maintain 60 percent of its land protected from development.

Source: Map from George and Vicky Ranney. Reproduced courtesy of Prairie Crossing.

▼ Plate 7   Prairie Crossing, IL

A learning farm associated with the Prairie Crossing school was incorporated into the community design to keep people connected to the land.

Source: Photograph from George and Vicky Ranney. Reproduced courtesy of Prairie Crossing.

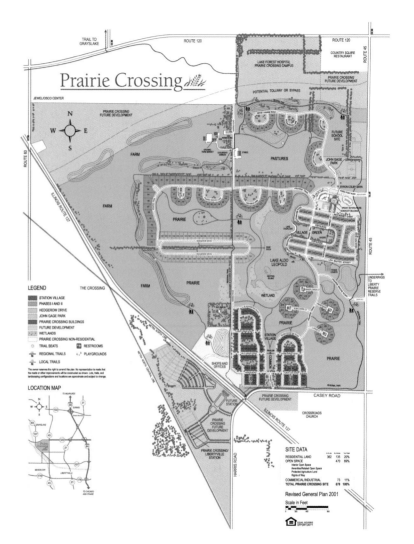

Housing built by the government, like this Rainbow Row development, can be attractive and should fit in with the surrounding architecture so residents feel they are a welcome part of the community.

Source: Photograph by Elana Navon, from www.charlestoncity.info. Used with the permission of the City of Charleston.

▼ **Plate 9   Charleston, SC**
Access to the waterfront is a gift Mayor Joseph Riley gave the people of Charleston. This unique decision in redeveloping the city encourages people to walk and to enjoy the beauty that surrounds them.

Source: Photograph by Joseph Sohm. Used with permission.

▲ Plate 10   Charleston, SC

Parks that provide shade and places to walk, sit, and gather become places for people to take exercise and to connect with their neighbors.

Source: Photograph by Elana Navon, from www.charlestoncity.info. Used with the permission of the City of Charleston.

◀ Plate 11   Elgin, IL

Walton Island, in the middle of the rejuvenated Fox River, is a popular site for events.

Source: Photograph from the City of Elgin, www.cityofelgin.org. Used with permission.

▶ **Plate 12   Elgin, IL**
In rejuvenating its community, Elgin chose to protect and restore its historical architecture.

Source: Photograph from the City of Elgin, www.cityofelgin.org. Used with permission.

▶ **Plate 14   Boulder, CO**
Pedestrian-only streets, with shade and places for people to relax, encourage healthy choices and build community.

Source: Photograph by Scott Harris, from Sangres.com. Used with the permission of Sangres.com.

▼ **Plate 13   Elgin, IL**
The fountain in Elgin's new Festival Park attracts children, who play in the water for hours. Once a superfund site, the park is a destination where families gather to enjoy music and picnics.

Source: Photograph from the City of Elgin, www.cityofelgin.org. Used with permission.

▶ **Plate 15 Boulder, CO**

Boulder uses a combination of on- and off-street bike paths to encourage bicycling as a safe transportation choice.

Source: Photograph by Kevin Krizek. Used with permission.

▶ **Plate 16 Oakland, CA**

One of the busiest ports in the United States, Oakland is making efforts to minimize emissions that affect the air quality and the health of the surrounding community.

Source: Photograph from the Media Policy Center.

▶ Plate 17   Oakland, CA

Multiple freeways have bifurcated Oakland and increased traffic in West Oakland; vehicle emissions are affecting the air quality and the overall health of those who live in proximity to these roads.

Source: Photograph from the Media Policy Center.

▶ Plate 18   Oakland, CA

This new development was built in proximity to the West Oakland BART (Bay Area Rapid Transit) station to give residents easy access to public transportation.

Source: Photograph from the Media Policy Center.

▶ Plate 19   Detroit, MI

An unfixed broken window can encourage further vandalism, leading to an abandoned building, and eventually a blighted community as people leave in search of work and welfare.

Source: Photograph from the Media Policy Center.

▶ **Plate 20   Detroit, MI**

Urban gardening repurposes blighted land into a source of healthy eating, and the bonds that grow between neighbors as they garden together are even more valuable.

Source: Photograph from the Media Policy Center.

▶ **Plate 21   Detroit, MI**

The Eastern Market, once a rail terminus for transporting goods through Detroit, has become a hub of local economic growth and a destination for local small businesses.

Source: Photograph from the Media Policy Center.

Planned communities, once thought to be a haven from urban strife, isolate both the young and the old because they cannot drive themselves and so have limited access to markets, school facilities, and friends.

Source: Photograph by Joseph Sohm. Used with permission.

Once a symbol of industrialization, the Beltline now links communities through walking and bicycle paths, connecting people and encouraging healthy choices.

Source: Photograph by Christopher T. Martin, www.christophertmartin.com. Used with permission.

▶ **Plate 24   Atlanta, GA**

When the anchor store of a mall closes due to changes in the economy or population decline, communities must decide how they will repurpose the land.

Source: Photograph from the Media Policy Center.

▶ **Plate 25   Los Angeles, CA**

The neighborhood coffee shop is an integral part of a community; people can meet for conversation and take a break from the pace of their day to enjoy the local surroundings.

Source: Photograph from the Media Policy Center.

▶ **Plate 26 Los Angeles, CA**
The automobile has become a symbol of Los Angeles. The highways act as arteries, moving cars throughout the city, but now they routinely become clogged, as they were never designed to handle the number of people who have flocked to this sunny, seaside community.

Source: Photograph by Joseph Sohm. Used with permission.

▲ Plate 27   New York City, NY
Frederick Law Olmsted knew that for a city to be healthy, its people needed places where they could walk and enjoy fresh air. The best of such places, such as Central Park in New York City, have become iconic to their communities.

Source: Photograph by Joseph Sohm. Used with permission.

Chapter 6

Saving America's Downtowns and Local History Through the Political Process

Charleston, South Carolina

A key way to improve the way people move, work, play, and stay healthy is to change a city's layout, its infrastructure. The buildings, roads, landscapes, and walkways affect how the community is experienced, and this experience influences people's behavior. Rethinking the structure of a community is daunting, but not impossible. Everything we look at today was in someone's mind at some point. We do not have to accept someone else's imagination; we can impose our own. Over the course of a century, the only things that really last are the road patterns and a few of the buildings, so choosing what to preserve and what to build reflects our priorities. It is through the political process that change occurs, and the results can be spectacular.

The city of Charleston, South Carolina, is a great example of what can be done over time with the right leadership. Charleston has a rich history, but a city cannot live only in its past. For a number of reasons, Charleston never fell victim to the destruction-construction disorder that afflicted the rest of America after World War II. In part because it was comparatively poor and not on the road to some other destination, it was spared many of the ravages of *urban renewal* (often called *urban removal*) and thus the "edifice complex" of *brutalist* buildings of plain concrete (of the sort I discussed in Chapter Three) and the coup de grâce of the interstate highway system roaring through an urban area.

This classic American city has preserved its downtown and local history, making it into a jewel of the region. As mayor of Charleston, Joseph "Joe" Riley has guided change in his city for more than a third of a century. Like all visionaries, he plans strategically for the long term and acts tactically in the immediate term. "I love what I do. I get to work early in the morning with great joy. It's the most wonderful job to work for this community and to have the opportunity every day to work on initiatives that can make the lives of our citizens better fifty and a hundred years from now," Mayor Riley says.

Just as good health doesn't just happen—it takes a daily effort to make good choices—good design and good places do not just happen—it takes political leadership, intelligence, creativity, and perseverance to turn difficult sites into great places that will support people's health. Mayor Riley has used the political process to make changes that can be sustained over the long term.

SYMPTOMS

Charleston's history shaped the city's plan, architecture, food, furnishings, language, cultural activities, and priorities. The first settlements in the area suffered bloody rivalries among the Spanish, French, British, and Native Americans, but also established a multicultural community from the beginning. When John Locke wrote the Constitution of Carolina, he guaranteed religious freedom, a principle that has influenced the social fabric of the city ever since. The arrival of slaves from Africa and invasions by pirates added layers of social complexity to the varied tapestry of the city's history. Involvement in the founding of America through political representation and armed conflict reflected a commitment to political ideals and action. Summer heat, earthquakes, and hurricanes have influenced architectural choices and the development of embankments and fortifications against weather

and seas. When Mayor Riley took charge in 1976, he was immediately faced with two critical issues, a dying downtown and a crumbling urban housing stock. Unlike the leaders in many other cities across the country, Riley committed the city to restoring its beloved downtown. He also fully understood the need to rehabilitate rather than tear down neglected but otherwise beautiful and affordable housing stock. As a result of his vision, the city of Charleston is today one of the most beautiful cities in America.

Creating Home Ownership

To create the Charleston that he imagined, Mayor Riley identified what the city's residents most needed, and that began with a home. The city of Charleston has been recognized as one of America's leaders in the creation of sustainable, affordable housing for its citizens, winning awards for design and best practices (skills that come together in the project shown in Plate 8). Riley believes that "if you let something bad be built (or have the government do something bad), it is repeated. If you do something good, it is also repeated."[1] The fight is at the *beginning* of the process. It's crucial to set a standard that, once successful, will be followed.

Mayor Riley is a co-founder of the Mayors' Institute on City Design, a National Endowment for the Arts leadership initiative in partnership with the American Architectural Foundation and the United States

Conference of Mayors. Since 1986, the Mayors' Institute has helped to transform communities through design by preparing mayors to be the chief urban designers of their cities. When Mayor Riley speaks to other mayors, he often explains that "we should not accept anything less than beautiful, properly designed . . . housing for low-income people. It is not acceptable to allow the [government] to build that disgusting, auto-oriented, HUD [Department of Housing and Urban Development] public housing (also known as 'ant hills'). We can refurbish old Victorian houses to provide affordable housing. It not only works well for the neighborhood, but is gorgeous as well." He goes on to say that these old houses have tremendous character and excellent design features.[2] Riley knows that replacing buildings that have character and history with low-quality, often dreary structures does little to inspire residents, so he has fought against having bulldozers entering redevelopment areas.

In this spirit, in addition to encouraging some well-designed new building, Riley has spearheaded initiatives to foster home repair, painting, and property rehabilitation for improving existing structures. Under his leadership, councils and nonprofits have formed to make housing accessible to families at different income levels. Charleston was designated in 1994 as an Enterprise Community. This initiative is a neighborhood revitalization strategy for improving affordable housing options and providing health, human services, and increased economic opportunities to neighborhoods that need support.

Creating Connectedness

When I talk with Joe Riley, it is like talking to a long-lost brother. What he says makes sense to me. We all know intuitively that he is right when he says, "People need to have public centers such as farmers' markets and public plazas and pedestrian malls. It is human nature to want to be with other people. We should not throw away these precious, centrally located, and accessible centers."

Riley talks about Charleston's parks and public spaces as his "generation's gift to the future—the birthright of the people of Charleston." (Plate 10 is an example of this birthright.) In fulfillment of this vision, he has pursued an ambitious plan to give the public access to nearly every foot of Charleston's precious waterfront. He has renovated many parks and playgrounds and used abandoned railroad rights-of-way for new bike paths and greenways. The open vistas to the sea and green spaces for recreation and relaxation, along with historic landmarks and buildings, create the *feeling* of Charleston, and such an essence is powerful when it is consistent. The challenge comes when developers want to create new spaces. "Almost always, when a timeless, gorgeous, dignified, historic building is demolished," Riley explains, "it is replaced with something that does not last long." The new building doesn't fit in with the surroundings, and people don't like it. In time, the initial business fails and moves out, leaving the unloved structure for others to deal with.

Part of what is important in maintaining a community is keeping people interacting. Most modern buildings promote privacy through the positioning of their windows and doors. But such isolation is not what Charleston is about. This is a city that has created a vision of people interacting, so as Riley explains, "Developers must be forced to design their buildings so that they are oriented to the street rather than turning their backs to the street, since it is so critical to build energy on the street—the street is an essential public realm."[3]

Creating Economic Development

"Small business is the economic backbone of our community," says Mayor Riley. He created the Charleston Citywide Local Development Council (LDC) to provide assistance and low-cost loans for businesses throughout the city and leveraged public funds with private development to enhance projects aimed toward expanding and growing business in the areas of tourism, shipping, medical services, and technology. Through partnerships, Riley has stimulated new development and restoration in the historic downtown, what he refers to as *the heart of the city of Charleston*. As his biography on the city's Web site says, "Mayor Riley has helped create one of the most vibrant and productive downtowns in America, including the dramatic rebirth of King Street, Charleston's main street."[4] Through partnerships and incentives, the city is attracting larger projects and international corporations to increase employment opportunity for all citizens.

Through much of the decade before the Great Recession that began in late 2008, Charleston's unemployment rate averaged 5.2 percent.[5] This is not easy work to accomplish. I'm sure that Mayor Riley's been in his share of political fights. To be mayor of a midsized, rough-and-tumble seacoast city for thirty-five years is no simple feat. But what makes the work remarkable is that he's optimistic, positive, and proud.

With economic development and rejuvenation of neighborhoods and open spaces, and also Mayor Riley's commitment to racial harmony and progress, Charleston is experiencing a substantial decrease in crime and an unprecedented growth in the city's size and population. Riley has led the city government in establishing an impressive record of innovation in public safety, housing, arts and culture, children's issues, and the creation of parks and other public spaces. Charleston is recognized as one of the most livable and progressive cities in the United States, and a steady stream of newcomers has boosted the region's population to nearly 650,000 (Figure 6.1). As the Charleston Web site puts it, "Mayor Riley has provided the leadership and incentives necessary to make Charleston a great place to work, live, and visit."[6]

DIAGNOSIS

Charleston is a crossroads of cultures, ethnicities, and economics, but rather than maintaining segregated

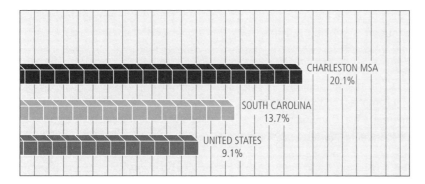

Figure 6.1 Population growth in the Charleston Metropolitan Statistical Area (MSA), 2000–2009.

Source: Charleston Regional Development Alliance, "Market Profile," Charleston Inspired (2010), http://www.crda.org/business/market_profile/. Used with permission.

Figure 6.2 Two-way streets allow involvement with the life of the city.

Source: Graphic by Scott Izen for the Media Policy Center.

neighborhoods, Mayor Riley has looked for ways to keep people in close proximity and to cross-pollinate people's ideas in order to encourage connectedness and to build community.

People with higher incomes demand higher-quality services. By maintaining both an economic and ethnic mix, people can teach each other about better ways to live and how to access services the city provides. The city's services should support the way people want to live—they should make it easier, healthier, and more enjoyable to live in the community. Even traffic patterns play a part in this. Charleston has chosen to have mostly two-way streets, even though some traffic engineers say one-way streets move cars not just more quickly but more safely. However, Charleston has found that people drive more slowly and thus more carefully and quietly on two-way streets, and perhaps they may enjoy life and the city more, particularly when those streets are lined with

parked cars, trees, sidewalks, interesting storefronts, and ideally, bike lanes (Figure 6.2).

Many traffic engineers have forgotten the reason we travel. We rarely do it for the fun of getting in a car (especially true in Los Angeles and Atlanta) but for being where we want to be—at work, with family and friends, or at the store. The purpose of transportation is to get where we want to go, not to move 3,000 pounds of machinery. We build quality density and *complete streets* so people can accomplish their tasks easily and safely.

Traffic fatality risks go up with the square of a vehicle's speed—compared to the death rate for people in a car crash at 25 mph, the death rate at 50 mph is not just twice as high but four times as high. Therefore, even though some found it surprising at first when Charleston decided to change Upper King Street back from a one-way to a two-way street, this change gave the street a more pedestrian feel and emphasized that it was a destination, not a mere thoroughfare. Dan Burden, senior urban designer and co-founder of Walkable Communities, describes a similar experience of changing from one-way back to two-way streets: "several decades ago the chief transportation planner in Washington, D.C., wanted to make the city people friendly. He oversaw a change of high-speed streets that were one-way back to two-way, and the city has become much more pedestrian friendly."

In years past many two-way streets in North American downtowns were changed to one-way in an effort to speed up traffic. It worked well in enabling suburban migration and commuting but led to the further demise of urban downtowns, walking, and social engagement. As we reclaim cities, many one-way streets are being studied and returned to two-way streets. As Mayor Riley says, "This is about improving the social life of our communities."

The position of mayor comes with legal and budgetary authority but also with substantial constraints. Change comes less from the mayor's power and more from his or her leadership and ability to leverage priorities and resources to encourage engagement within and beyond the government's direct influence. When visioning the redevelopment of downtown Charleston, Mayor Riley had to look at both the residential and commercial potential of the community. Decisions about industry and economy are vital. Will there be chain stores and restaurants, small businesses, manufacturers, or intellectual enterprises? Knowledge-based businesses, for example, need cutting-edge technology, spaces of various sizes, and manageable costs. They need an environment that inspires creativity without getting in the way of efficiency. But most of all, knowledge-based businesses require creative young people—these individuals are the engine of future growth and they do not want to live in boring places. They want liveliness and a good place to raise families. And they do not want to emulate the long hours their parents spent sitting in traffic.

Minimizing the Footprint of New Buildings

Leaders and citizens in downtown Sacramento, California, are exploring many of the same ideas that Mayor Riley has worked with in Charleston. They are trying to retain all the components of historic buildings that they can. Nothing is greener than leaving what's there in place, but if an old building is loaded with asbestos and toxics, it may need to be carefully demolished. It takes a lot of thought to rehab a historic building, and sometimes it's not possible.

When we modernize existing structures, we owe it to our future to use green building practices and sustainable infrastructure. In Charleston, to keep construction costs manageable and prices and rents reasonable, some of the old buildings have been retained to lend character, but they have also been adapted to handle the technology demands of creative entrepreneurs. Such *adapting and reusing* is part of creating a livable and thriving community.

Surface parking suffocates cities—we have all seen the places that look like Monopoly boards, with random high-rises surrounded by vacant lots doubling as car parks. In a sunny climate, parking lots, especially black-topped ones that create even more heat, make walking miserable and can make a city street feel like a no-man's-land. Parking lots generate trivial revenue; the owner's goal is to hang on to the land long enough for this "real estate" to become a good investment, say as a lot for building a high-rise. If we must have parking facilities, stack them (Figure 6.3), even if that is more expensive. This then allows us to turn idled asphalt lots into community gardens that will feed us and help to cool the environment. Best of all is to put in transit to make it easy for people to get to work and to save money, and also to put in much more working-class housing so the downtown is alive 24/7.

How many people bother to take pictures of 90 percent of what we've built in America over the last sixty years? Yet when people walk into a refurbished trolley carbarn or an old school, they smile as they take out their cameras. This is about allowing people to feel good about where they live and work. Maybe architects are thrilled, but I do not see much excitement in most people about brand-new buildings. When an old building is updated and goes back into use, a part of the city's history returns. When people are excited about what the city or another developer is doing, it makes redevelopment a whole lot easier. People take a little bit more pride in where they work, and they want to show people what's happened to the space. The space becomes a part of the conversation, a part of the culture.

Figure 6.3 A parking structure that pedestrians may not immediately recognize as a car park.

Source: Photograph from the Media Policy Center.

CURE

Look around Charleston and you can see who won in the preservation of historic downtown. There were two battles that Mayor Riley had to face—one for design and one for density. Riley believes "cities need thirty-story buildings downtown like they need a hole in the head." As he explains, five- and six-story buildings are much better, because they define public spaces and do not force a city to have huge expanses of parking. Most cities in the United States look as though a neutron bomb had been detonated. Vast expanses of baking-hot parking lots, absolute no-man's-lands, flourish between high-rises. The vista is about as erosive of the soul as anything we could imagine.

This goes back to the importance of planning and city codes. "Infill is critically important to the restoration of cities in America, but the infill needs to be at the right scale. It needs to fit into the neighborhood's DNA, not something that was borrowed from somewhere else. The building has to have the right form, use the right materials, and reflect Charleston's rhythm," says Mayor Riley. The area surrounding the building has to reflect the culture as well. The landscaping and color choices should reflect a sense of place. (Plate 9 shows an example of Charleston's coastal architectural design.)

Architecture has eras, some critics call them fads. So, for example, the neoclassical style of Washington, D.C.'s iconic buildings gave way to the rococo stone pile of the Old Executive Office Building and then the style shifted again to the modernist starkness of the FBI building and now to postmodernist structures. These styles usually reflect similar changes in fashions in clothing, literature, and art. Every building starts with an intended function. Moreover, buildings must be *legible*—a visitor should be able to read essentials like access, egress, and destination just from the design—and buildings must also convey identity. Is it a court or city hall? It should project authority. Is it a hospital? It should project a caring environment through sober serenity. Is it a school? It should welcome with order, calm, and joy.

The debate between form and function is perhaps akin to the battles between any two things that are only partly realized when they lack each other. For example, the symbol representing the Tao has two parts, yin and yang—each represented as a loop with beauty of its own, sleek but incomplete. But when these two parts are combined they achieve perfection—the circle. I think Charleston has found its Tao. It has commerce and business, and it has a delightful culinary community, arts, music, and public social events. "There should be energy on the street, and buildings should contribute to that. Parking garages don't need to look like garages. They can look like office buildings and have retail on the first floor. Downtown should be designed for people," says Mayor Riley. "It's disrespectful to build auto-oriented facilities that look like auto-oriented facilities, so I stuck to my guns and the architects usually gave in. They often decided afterwards that the feature I insisted on was the best thing they did with the project."

A Charleston Resident Comments on Social Diversity and the Housing Market at a Neighborhood Town Hall Meeting Led by Mayor Riley

I live on Smith Street right across from a public housing project. This is half the reason I like this neighborhood so much. Some of the first developments were done right around the corner, when this wasn't a dilapidated neighborhood but it was nothing like it is now. I like that it isn't completely gentrified. I like that we have permanently affordable housing and subsidized housing across the street from the market. We all know that neighborhoods that are diverse are healthier so it's something we work to protect. We still have a bunch of African Americans and nice houses. Property values have been rising so this kind of development has helped everyone.

Across the country there has been a downturn in housing prices, but here, in an area that traditionally hasn't done very well, we're not experiencing this downturn. We're sustaining our housing prices better than South Florida or Southern California. Houses may be staying on the market a bit longer than they used to, but they're still up in value a little bit. They haven't even dipped very much in downtown Charleston, so that's very special. There's only so much peninsula in Charleston, so the demand is holding.

This goes back to the prior discussion of monoculture versus ecosystems. Everyone knows that higher-income people need lower-income people sooner or later, for help with preparing meals, maintaining homes, teaching their children, caring for their loved ones, and patrolling their streets. But poorer people need richer people too. The tax base and influence of the latter draw to the neighborhood better security and services. And an economic mix is good in schools too; richer and poorer kids need to learn together at school and to learn from each other.

Mayor Riley's overall take on downtown Charleston is about managing development to fit the vision:

A city's downtown is like an ecosystem. When I was a child this was a shopping district put in for this part of our state, and then like most main streets, we were hit hard. The new distant shopping centers drew people away from the city. We were determined to restore King Street because every great city in the world, whether it's little or huge, has a healthy downtown. It's a public ground, a place that people own. They know this is where they celebrate their citizenship, the diversity, the rich, the poor, the young, the old, the shops, the energy, the elbow contact, the eye contact—the pulse.

Downtown is the energy of the city, its heart and soul. To revitalize downtown you've got to go back to the basics. Create a vision with the stakeholders by asking the critical questions, devise a comprehensive plan based on the vision, and implement the plan. We did a thorough strategic plan to make sure we knew exactly what we needed to put where and then went about doing it, never compromising with the plan or with quality.

As we were working we had to be careful that we put the right thing in the right place. It's got to be the perfect use with the perfect design in the right place to give the restorative energy to the ecosystem. That's what downtowns are—they are ecosystems, so if you put the right agent in the ecosystem then it is restored to its former life.

Mayor Riley is not just a politician. He is a visionary. When we think about public space and public health, we have got to look at where we put urban density. I believe the most important land should be for the public and the future. The buildings and the physical environments we create in the public domain should inspire people and they should be beautiful places, because psychological health is physical health.

PREVENTION

Changing a Culture

Charleston has a long history of racial tension, and that is something the city tries to address through education, housing initiatives, and public forums. *Southern Living* magazine interviewed Mayor Riley in March 2010, asking him about his most indelible image of the city's changing culture, and his answer says much about where the city is on its struggle toward racial equity:

> Once the Ku Klux Klan requested a permit to march here. We had to give them a parade permit and we were pretty nervous—all kind of bad things can happen at one of their rallies. Reuben Greenberg, our

police chief at the time, was both African American and Jewish. He was wonderful. He told the Klan he would lead the march to protect them. I watched as this brilliant, courageous man led the Klan parade down Meeting Street. There was no trouble. That's something I'll never forget.[7]

Although the implementation of the permit was not what the Klan had envisioned, the city treated this group's permit request as it would have any other and provided the protection that was needed to ensure the peace. It is one thing to have a vision of a community that respects free speech but quite another to follow it with confidence and humor when faced with a difficult and delicate situation.

Seeking Architects and Contractors Who Can Be Friends

Throughout our multispecialized society, we have lots of rules—codes for buildings, requirements for school acreage, trade practices, and transportation plans. I once had an architect say to me, "Contractors want to cut corners, and they secretly just want to be architects." I once had a contractor say to me, "Architects don't know anything about building. They have never had to make hour-to-hour adjustments, in part because plans are never perfect."

It is no accident that the father of architecture, Vitruvius, was an architect and a builder. He knew that good buildings started with people's needs, and that for practical reasons, most were made of local stone or wood. By starting with people's needs, we ensure that the scale is comfortable. Moreover, a really *good* building has to be organic, in the best sense. It must fit *in place* and *in time*. And this fitness is not static, because the culture and the people change, along with their needs and ideas.

Builders must look at many sets of guidelines, as well as their balance sheet and return on investment. But the critical measure is the effect on people. Does the proposed structure work for people's health today and in the future? The structure they create will affect the built environment not only for the immediate user but also for all the future users. As Mayor Riley attests, the built environment is a legacy in concrete.

Chapter 7

Reinventing a Healthy City Through Community Leadership for Sustainability

Elgin, Illinois

Elgin, Illinois, is an American city of about 100,000 people, founded in the 1830s as a stop between Chicago and the new cities springing up throughout the Midwest, and it grew into a center for industry. The Elgin National Watch Company became the city's anchor employer in the 1870s, employing 500 men and women when the factory opened in 1867 and burgeoning to nearly 3,500 workers in the late 1950s, becoming the largest producer of watches in the United States. The company closed its doors in 1960, and the factory was demolished in 1965. This makes me think of how the United States has changed during my lifetime. When I was a boy in Newark, New Jersey, there was a tall, old-fashioned Elgin clock, with big numbers, in front of a jewelry store downtown. As a child learning to tell time, I was fascinated by it. It had a verdigris copper stanchion and must have been fourteen feet high. I can

imagine it even today, and now, like the factory and a great deal of other fine manufacturing in America, it is gone.

Yet Elgin has once again become one of the fastest-growing cities in Illinois, largely as a bedroom community about forty miles west of Chicago and about fourteen miles north of Fermi National Accelerator Laboratory. The city plans to add 14,000 homes and an additional 60,000 people by 2030. However, Elgin has been facing problems similar to those of many communities nationwide—an environment built (or modified) for cars, the resulting suburban sprawl, and a lifestyle that promotes obesity (Figure 7.1) and poor health. In 2009, through data from the Centers for Disease Control and Prevention and a survey administered through driver's license renewals to self-selected participants, Elgin was identified as the most overweight community in the state of Illinois, with

Figure 7.1 Weight gain is a national problem.

Source: Graphic by Scott Izen for the Media Policy Center.

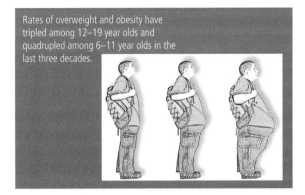

Rates of overweight and obesity have tripled among 12–19 year olds and quadrupled among 6–11 year olds in the last three decades.

63 percent of adults and 16 percent of children found to be overweight or obese.[1]

The people of Elgin are working to improve their city, and the city has created the Elgin Sustainability Master Plan to manage its growth. Elgin's former mayor, Ed Schock, summarized the local movement as a "citizenry ahead of the elected leaders. The citizens have been on the leading edge of the sustainability movement, and government is

trying to catch up. The one thing I do know is that building sustainability will be the longest-standing effort that the city has ever convened. This is not something that we're going to complete in a year or two, or even in a decade—it may never be completed, it may be something that's impossible to complete. This is an ongoing effort."

SYMPTOMS

Elgin was a classic example of an unsustainable community, economically and environmentally. Just as your personal financial portfolio should be diversified, so also should a city's revenue streams. In the mid-1950s, when the Elgin National Watch Company employed 3,500 people, the population of the city of Elgin was 45,000. In essence, Elgin was a modified company town (Figure 7.2). In the early 1960s, the watch business started to decline, due to the arrival of quartz watches and international competition. The company moved a portion of

Figure 7.2 The home of the Elgin National Watch Company in the 1950s.

Source: Photograph from the Elgin Historical Society. Used by permission.

the business to South Carolina at first, but by 1965 it had closed its doors and the factory was demolished. Almost all the families in Elgin were supported by a single wage earner, just like most families across the United States at that time. That job loss for 3,500 workers meant that 13,000 people in Elgin—almost one-third of the city—lost their income. Beyond the direct loss to factory employees and their families, the services these employees used, like the car dealers who had built their lots outside the factory gates and the markets and all the small businesses of Elgin, also were facing a loss. Pretty soon, 50 to 60 percent of the Elgin population was financially affected and the economy collapsed. The whole community changed.

With the loss of the factory, Elgin's social support systems became overwhelmed. Those who contributed to social programs through taxes and contributions had little left to give. Instead of supporting these services, they themselves became dependent on them, swelling the need beyond what could be absorbed by a declining budget. Elgin fell on hard times. The city was a monoculture, and the loss of the main crop, watchmaking, meant collapse.

Once businesses closed, buildings deteriorated and the health of the community declined. Dependence on a single source of income had weakened the community. It needed a new vision, and that vision was *sustainability*.

Dave Kaptain, then city councilman in Elgin and now mayor, is working with the schools and the community at large to implement Elgin's Sustainability Master Plan. He says: "The great law of the Iroquois Confederacy captured sustainability as the impact of our decisions on the next seven generations. I'm not as eloquent as the people who wrote that. I look at sustainability as a three-legged stool—one leg is the environment, another is economic development, and the third is social equity and equality. Sustainability means that the community has to grow in balance so that all three of these legs grow at the same time. If you can't do that, the stool gets tippy to the point that it falls over" (Figure 7.3).

Figure 7.3 The three-legged stool: the pillars of sustainability must remain in balance.

Source: Graphic by Scott Izen for the Media Policy Center.

Students Make a Medical Diagnosis: A Bad Built Environment

Yeng and Ron, students at Elgin High School, participated in a community event marking the culmination of a research project carried out by an environmental studies class, led by teacher Deb Perryman, that addressed sustainability and community health topics. Their presentation, which looked at the obesity study mentioned earlier, raised some issues for the residents of Elgin to consider deeply. They stated:

> Based on a study conducted by the Illinois Department of Motor Vehicles in June 2008, not only is Elgin the eighth largest city in Illinois, but it is also the fattest. This is definitely a concern for us because we are worried about our family members who have unhealthy habits and how they are influencing the younger generation. We are worried that becoming overweight at a younger age will result in physical or mental health problems, impact their self-esteem, and their drive to succeed in school. A lack of confidence can keep them from experiencing opportunities that will help them advance in life. Approximately 30 percent of Illinois children are obese or overweight. Nearly 30 percent of those twenty years and older are taking high blood pressure medication or have hypertension. That's really scary.

Dr. Jeff Bohmer practices medicine at the St. Charles Tri-City Health Partnership Clinic in Elgin. He sees how obesity contributes a great deal to diseases the clinic is now treating on a routine basis, including hypertension, diabetes, and orthopedic issues:

> Back and knee pain are very common complaints at the clinic. The more weight a person carries the more pressure is exerted on the joints and spine. Not only is a person's well-being affected by these ailments, but so is our society. When [people are] injured or sick,

treatments raise costs and these people cannot complete their duties at work, and business productivity is hurt. I think our local government needs to consider how we are going to encourage people to get to the store, their local train station, or get daily tasks accomplished, without using a car, then make the changes to our city to make this a real option.

DIAGNOSIS

The Elements of Sustainability

To re-create a thriving community, Elgin needs the elements that Dave Kaptain described—a healthy economy, a good environment, and a measure of social equality. The choices the city makes should create a diversified local economy that can survive in the present without limiting or bankrupting the future. These choices should promote a healthy environment, implementing practices that not only leave a minimal footprint on the land, water, and air but also improve these aspects of the natural environment. Elgin needs to retain and attract creative and energetic people who want to lead successful and healthy lives. What is important is that everyone becomes involved—the City of Elgin cannot do it alone. Businesses cannot do it in isolation, and individuals cannot do it by themselves either. When everyone works together, a big impact can be made.

Kaptain explains how Elgin is "trying to lead by example. Our Sustainability Master Plan will guide our community for the next forty to fifty years. Most of the people we run into think we are just talking about compact fluorescent light bulbs. Even though that is something, there are other things that are very important to creating a sustainable community." While serving for six years on the Elgin Planning Commission, Kaptain reviewed annexations and new businesses wanting to come to Elgin. "I started to appreciate the value of economic development and how that revenue, and the businesses and people that are associated with them, help form the basis for a

community. We're still working on that part and probably will be for the rest of my life."

Elgin is in Kane County. Each year the county tracks progress on its identified health priorities and works toward its identified goals. In 2011, the county planning department asked me to give the keynote address at its 250-person meeting on *smart growth*. This was the first time the county had included a public health component in its master land-use plan. County leaders have come to realize that land-use decisions can have a "monumental impact on lifestyles, such as the inclusion of *walkable* neighborhoods, access to fresh fruits and vegetables, and open spaces for active living."[2] As I have been discussing throughout, we know that if cities construct environments that make it difficult or impossible for people to walk and thereby remove this incidental exercise from people's lives, then people's level of fitness is reduced. This also drives away the lively creative people that cities need to thrive. Cities can create environments that can be much smarter in terms of protecting the planet, protecting human well-being, and in the long run, protecting the economy and prosperity.

I think the sustainability movement is closely related to personal health. We have an obesity and diabetes crisis, and we have not focused on prevention. This isn't just my opinion. The Kane County 2009 Annual Report, for example, shows that diabetes hospitalization in the county has increased since 2002 from about 115 to 151.9 per 100,000 (Figure 7.4). The percentage of adults who are overweight or obese has started to decline, but

Figure 7.4 Diabetes hospitalization 2002–2008: rate per 100,000 population.

Source: Kane County Health Department, *Vital Signs: 2009 Report to the Community*, Annual Report (n.d.), p. 10, http://www.kanehealth.com/PDFs/AnnualReports/KCHDAnnualReport09.pdf. Used with permission.

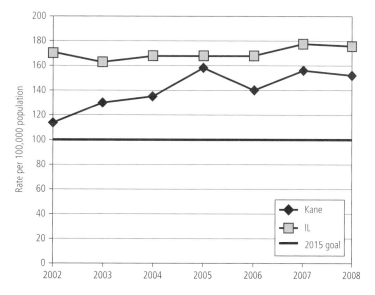

Figure 7.5 Adult overweight and obesity 2002–2008: percentage of adult population.

Note: "Collar" refers to the suburban counties surrounding Chicago.

Source: Kane County Health Department, *Vital Signs: 2009 Report to the Community*, Annual Report (n.d.), p. 15, http://www.kanehealth.com/PDFs/AnnualReports/KCHDAnnualReport09.pdf. Used with permission.

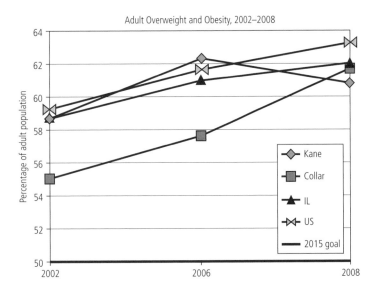

Adult Overweight and Obesity, 2002–2008

at 60.8 percent it is still well above the goal of 50 percent by 2015 (Figure 7.5).

One way I think we can enhance public health is to build facilities, amenities, and natural places so people can get out of their home living or recreation rooms, away from the television or electronic game, and make use of bicycle and walking paths, engage in a workout in the company of other people at a recreation center, or take the dog for a walk along the river. Exercise doesn't have to be something that becomes an ordeal. I think investment in public facilities will cause people to get out and get exercise that they can enjoy.

When people interact, they improve their mental health and their community's health. Talking together, walking, and bicycling are part of increasing community health and solving the health care crisis. We focus very much on disease treatment and nowhere near enough on prevention.

The City of Elgin, leveraging its own capital investments with much larger investments from the private sector, has invested a lot of money in its recreation center and has created a Bikeway Master Plan to make people more bike- and pedestrian-oriented. A group called Activate Elgin is working to bring a healthy diet and healthy conditions to children. Those involved in the sustainability effort know that it's very important to get across to the community that *a vital part of sustainability is being healthy*. From a profitability and an economic development standpoint, businesses have started to learn that it's important for their employees to be healthy. Good health adds millions of dollars to the revenue stream of the community. (Plate 11 is an example of designing for healthy conditions and shows part of Elgin's Fox River bikeway.)

The Impact of Sprawl

Zoning establishes the ways that land can be used in particular areas, and it can have a positive or a negative impact on our communities. Zoning has done good things—no one should have to live next door to a smoke-spewing, litter-creating, drive-through, fast-food restaurant—but zoning has also had unintended and undesirable effects. Antiquated zoning laws have divided neighborhoods by dictating sprawling single-family tract developments with minimum acreage requirements, ignoring the wishes of single adults or the needs of elders living by themselves. Such laws have also segregated citizens by economic levels. When zoning codes require large grocery stores to be located on parcels of a certain size in order to provide extensive parking, a much-needed grocery store—offering fresh and healthful food—may not be allowed to locate in a dense, urban area, even one where many people are within walking distance. This can lead to *food deserts*; many folks in poorer neighborhoods have no grocery stores offering a range of healthy foods in their neighborhoods.

One way to fix our zoning laws is to encourage mixed use of housing and low-impact commercial enterprises and to require that developers provide sidewalks, transit stops, and other active transportation facilities. Zoning laws should prohibit fast-food restaurants from being placed within a mile of a school.

Charrette is a charming French word originally meaning simply "cart." A charrette was used to collect the drawings of architects and their students in the Beaux-Arts school for judging events. From that context it came to be applied to a coming together of many designers, and today it often means a larger planning event for a project, one that can bring in not just architects but also planners, other specialists, and community members. I know from my own experience in planning both an office building and a large laboratory at the Centers for Disease Control and Prevention just how important it is to have the people who will be working in the building, especially in a laboratory where they will have specific workspace and utility needs, be involved in the overall building design. Charrettes enable citizens to talk to city planners and get a sense of the existing zoning codes and of the ways zoning policy might be changed to accommodate their own wants and needs and to best suit the wishes and needs of the community.

These changes coming from grassroots will typically have considered the health impacts of these decisions. Too often, design decisions are narrowly focused and sometimes have deleterious effects on the broader community. City departments of health, parks, transportation, planning, education, and police have not interacted in the past around such decisions. Infrastructure elements such as sidewalks are often still seen as amenities and not as transportation facilities. When community stakeholders start understanding and expressing the multitude of needs in a community—noting, for example, that one-third of American households do not own cars—the policymakers begin to understand that range of needs—realizing, for example, that providing infrastructure only for cars does not serve all people in a community.

Suburban developments took diversity out of neighborhoods. Gated communities of 1,200 homes reflected age restrictions, targeted ideal household sizes, and appealed to residents within limited economic parameters. Such developments, whatever their particular restrictions, are essentially monocultures. They often have two- and three-car garages and are designed to be car dependent. People who cannot drive, including teens and the elderly, have no way of getting around. They are dependent on those with cars or they are isolated. In older cities not originally designed for cars, some people feel that wide streets reflect progress whereas others want high density and buses or trolleys to get around. A city needs a vision that encompasses the needs of the many.

From a sustainability standpoint, age, income, and social diversity in a community are important, and the housing stock must reflect this. A young person or a young family wanting to make a start in the community needs less costly housing options and lower external costs, for things like transportation, in order to make a start. All communities need young families to survive. In-town rental housing is important to help them get a start.

CURE

One of the things Elgin decided to do as it began the process of change was to clean up its natural environment. Next, the city began designing sustainable neighborhoods to meet the needs of those who live and work there. The City of Elgin now has nine working groups dealing with revitalization issues: transportation, urban design, water resources, parks and green infrastructure, recycling and waste management, green buildings and technology, alternative energy, economic development, and education and outreach. City council members and Mayor Dave Kaptain are working on ideas coming from each of these groups.

Creating Cleaner Water, Land, and Air

Elgin faced real challenges in its push toward sustainability, starting with the water and the land around it. Upstream of the City of Elgin, the Fox River has long been dammed for power generation, has been used as a water source for dairy farms, and has been contaminated by pollution from paper milling. Over the history of the river, dredging has deepened it for commerce, but the sediment on the riverbed was contaminated from the pollution and removing that sediment transferred the contaminants to the nearby topsoil, which affected farming near the river.

In early 1970s, the Federal Clean Water Act required U.S. cities and towns to begin cleaning up and protecting natural environments, with a goal that all waterways become swimmable and fishable. As a result of the billions of dollars spent on cleaning up the Fox River over the past thirty years, Elgin has cleaner drinking water.

A huge side benefit was that the waterway became more desirable for nearby land-based activities as well as those on the water. If the Clean Water Act is ever revised, I suggest that it should require all urban and suburban waterways across the United States to be walkable, perhaps not always right along the riverbank but within a reasonable distance of it.

In addition to the problems with the river itself, industrial dumping had contaminated the property around the Fox River. Even the property where the public library sits required over a million dollars worth of remediation. To the north of Elgin is a twenty-acre Superfund site, a place that was once home to almost every imaginable toxic chemical. Through investment and a lot of work, the river areas have been cleaned and are now suitable for development. Costs and regulations scare private developers away from environmental cleanup, but once access to a desirable area around a river, or other natural feature, is restored, they will build facilities and amenities that draw people to the riverfront for work, play, and leisure. (In Plate 13 Elgin residents gather to welcome spring.)

Employing Mixed Use to Combat Sprawl

Elgin is fortunate to have great medical facilities. Sherman Hospital goes back to the 1800s and has been through many expansions. As the city thinks about sustainability, it is working on ways for Sherman and St. Joseph Hospitals to have employees live nearby. It is an issue for these hospitals to get people with the skills needed for the level of compensation they can afford. Providing adequate housing for staff could solve a number of problems. Some nurses commute fifty miles a day to get to work because they cannot afford to live near the hospital they serve.

The Grove is a relatively new development of upscale restaurants and hotels surrounded by an industrial park. An important next step is to create apartment and rental properties for people who work here. The best places to live and work are lively eighteen hours a day, not just eight.

River Park Place, also one of Elgin's newer developments, was conceptualized to revitalize downtown by providing living space for people who commute to Chicago. I think the vision needs to be broadened because people should not have to leave one downtown to work in another. The development needs more retail space and more low-cost housing, so employees can live where they can walk to work. This idea of proximity rejuvenates an old purpose for this downtown area, one from over one hundred years ago.

Living near work isn't a new idea. The Elgin National Watch Company built housing around its factory. There was a hotel across the street with men's and women's quarters, a dining area, and entertainment, and there was a train station. New employees were afforded low-cost rental housing. The train brought commuters from the outlying areas to and from work and took employees living in the subsidized housing to places of interest when they had leisure time.

More recently, the concept of corporate housing has evolved and employs many of the successful elements of the original mixed-use concept. The county environmental department is currently reviewing Elgin's zoning laws to see what can be changed to support the development of mixed-use sustainable communities.

Creating Sustainable Suburban Subdivisions

West Point Gardens, a subdivision within the City of Elgin, is under development by Rick Randle. His vision is an open space mixed design that offers privacy as well as community. The twelve-acre common areas are green, planted with flowering trees, and designed to encourage people to enjoy being outside. There are bike paths and sidewalks along the banks of a wide pond, and fountains here and there.

Residences are tiered at various prices. There are townhomes mixed in with smaller and larger single-family homes. Homes are designed to have wide front porches so people will sit outside to enjoy the sunsets in the summer, the views in the winter, and talk to their neighbors. Backyards are enclosed areas perfect for a vegetable garden, a fire pit, or other design that interests the owner. Garages are in the rear of the units so their blank doors are not visible from the street. As Randle explains, "We want to encourage people to walk, talk with each other, and enjoy the outside."

Diversifying Downtown Housing

No community is sustainable with one type of housing. We have to provide housing for all residents, and downtowns are among the good places to do this. If we think of downtown as representing the full spectrum of our community, we can begin to see the many kinds of support the built environment will need to offer. People have a variety of needs. Townhouses, condominiums, apartments, and single-family houses can all be mixed within a downtown area. Often the original buildings or townhomes in downtown areas will have been built with a number of stairs, and that's not something that a lot of older people find attractive. By age sixty-three, people's knees aren't working as they used to. We need to think about housing that is realistic for seniors, perhaps places to rent.

As people living in the suburbs age, they eventually develop transportation needs that cannot be met. A downtown area can provide them with housing, social contact, and entertainment; Elgin's downtown has a public library and a concert hall, for example. Seniors are usually on fixed incomes. Rather than going to the typical assisted living, what if seniors living independently were given vouchers to eat at the neighboring restaurants? This would also get them out walking and enjoying the fresh air. Restaurants have a quiet period between the lunch rush and the after-work cocktail crowd. If meals were provided to seniors during this time of day, restaurants might be able to give more people full-time work.

In the beginning, Elgin's leaders considered developing only upscale housing in downtown Elgin, so that the higher disposable incomes would support downtown businesses. But many average restaurant workers cannot afford to live in downtown Elgin, so the city needs to provide a housing subsidy for them. We need to provide housing for a variety of people in a variety of situations if we want our downtowns to be a reflection of the greater community. We need to think about entertainment and transportation, schools and parks. (Plate 12 shows one of Elgin's Victorian treasures that has been renovated and maintained.)

Tapping the Power of Designed Open Spaces

Festival Park in downtown Elgin was once a contaminated site, known as auto dealers row. As employees left the watch factory they would walk past the car dealers to get to the buses and trains. When the watch factory went out of business, the car dealers were left without customers, and after the dealerships folded, underground storage tanks for oil and fuel remained on these sites, sometimes for decades.

When Elgin decided to renovate the area, creating Festival Park (Plate 13), it first needed to deal with the underground tanks. The tanks were excavated and bacteria were used to remove the gasoline and oil residues left in the soil. The city brought in a casino to occupy one end of Festival Park and in the center of the park is a fountain

that young people use by the hundreds in the summer time. Parents sit along the benches that rim the fountain. Dave Kaptain reports that "it was beyond our expectations as to what impact this would have on the community. I thought that it would be an attraction, but nothing like what it has turned out to be."

The Bikeway Master Plan will eventually connect all the neighborhoods in the city. From the park, residents can ride their bicycles through Wisconsin, to the city of Aurora, all the way to Chicago, or out to Iowa. The part of the bike path that runs between Festival Park and the river engages people with the river and allows them to enjoy the benches along the shore. Fishermen use the benches during the summer, making this one of the city's most popular parks. The park has unexpectedly changed the downtown community. Row houses and townhomes that have been built alongside the river have been a positive development for people who live in the area—old-timers recall that this site was once contaminated and had been vacant for twenty years or so. Now the city council has committed over $30 million to redevelop the land—to make a beautiful site that allows people to exercise and to enjoy the new recreation center, the restored riverbank, and the cleaned-up environment.

Walton Island (Plate 11), a little bit downstream, is now a beautiful site where people like to have weddings. Former Elgin mayor Ed Schock recalls explaining to the city council, at a time before he was on the council, that "Elgin was a community where the only way you could see the river was if you stood on one of the bridges because there was no public access. An extra benefit for fixing up the riverfront as a public facility, putting a lot of our public amenities like the library and recreation center, along the river, is that they spur private investment."

"For every dollar the city has spent," Schock explains, "we have leveraged that into two or three dollars of private investment through new kinds of buildings, row houses, and businesses that have opened because the river has a magnetic quality." People want to be near water. Once you make the experience a good one, you entice private investment. People see the opportunities that walking and biking paths have to offer. It is no accident that almost every new development that's occurred in downtown in the last ten years has been along the river.

As Elgin grows, the city needs to think about creating more parks and more recreational open space for people. The city council has discussed shrinking the sizes of properties, making buildings a little bit denser, and creating larger and more numerous open spaces.

Creating Sustainable Businesses

There are multiple facets to a sustainable business—what is produced, where and how it is produced, and then what happens to the product once it is no longer needed.

Janet Jarecki owns Rieke Office Interiors, a business that manufactures office furniture and also helps clients create environmentally responsible workspaces. "Our new business space," she says, "is twice the size of our old place but our heating bill is one-fourth of what it was before. If we build more buildings that are green and very efficient, the people working inside will benefit, as well as the community. Using local materials in buildings reduces the environmental impacts of transportation and keeps people close to home working. When we support local businesses, more people want to work for us, and our business benefits as well."

This is an example of an organic business. Businesses are successful when they are the right industry in the right place and time. When Janet Jarecki designs environmentally conscious workspaces for her own business, she can prove that certain technologies work and can sell them to her clients. "We have switched to energy-efficient lighting with sensors that automatically dim the lights when there's lots of natural sunlight and increases the lighting when it is needed," she explains. "The building is well-insulated, keeping heat inside in the winter and the cold outside, which is great. We recycle all of our materials—we try to be environmentally conscious and aware."

On the product side, Jarecki does things to make her product more sustainable and better for the environment. "We manufacture office furniture. We use all water-based adhesives and the particleboard substrate is 100 percent postconsumer product. We use materials in our flooring and our fabrics that have a high-recycled content."

When I look around my office at UCLA I can see some of the ideas Janet Jarecki describes. For example, some offices and institutions are taking part in carpet rentals. According to Carpet America Recovery Effort's 2007 Annual Report, over six billion pounds of solid waste in the United States is used carpet. Much of this carpet was originally made from fossil fuel–based plastics and is highly recyclable.[3] So smart businesses are installing carpeting that they use for five or more years and then return to the manufacturer, or in many cases the owner, to be reprocessed, often into new carpets. Similarly, Jarecki's business increases the life cycle of existing furniture:

> We give our clients options for refurbished filling of existing office furniture. If one company goes out of business or is moving and upgrading, we'll buy back their furniture and sell it at a reduced price to a client who may be starting up or be on a very tight budget. By reusing furniture between facilities we can help our clients keep their costs down and keep furniture out of the landfills. We have a flooring department that not only lays down floors, but they help clients recycle their old flooring materials that are being removed. We think about the whole cycle from raw materials through use and to dismantling.

Another example of such efforts among Elgin's businesses comes from Elgin high school student Ashley Lundgren, who interviewed local business owners on creating sustainable small businesses for a survey that was part of the same sustainability and community health classroom research project mentioned previously. Here's what one of her interviewees told her:

> Since we opened our deli a year ago, I've been slowly making changes that reflect our environmental priorities, but these changes are also good for business in general. We replace our incandescent bulbs with compact fluorescents. I have started to purchase compostable green cups made of cornstarch, so even if they're thrown away they'll decompose in the landfill. We're switching our "to go" packaging to minimize the use of plastics. For those who dine in, we are using plates and porcelain mugs so we're not throwing away so much material. We also recycle.

> I have my own garden out back, which helps sustain me a little bit. I can't obviously provide all the food for my business with that garden, but I can do a good portion of it through the summer months. I do believe that as it becomes more difficult to get local produce, farmers are hurt. There is a direct impact, you know. The more we become urban, the more things have to be transported to us, and the jets have to fly, and the trains have to haul in products, so the greater our carbon footprint. I think it has an impact on public health on a great scale. We might only be making small changes where we can, but over time, it adds up and we will make an impact. Breaking our dependence on fossil fuels can have a healthful and sustaining impact on all of us, especially with gas now at close to $4 a gallon.

PREVENTION

In Elgin, success is measured by the people who are involved—the idea is that everyone should be engaged in the process from the perspective of how he or she lives, works, commutes, and plays. Bringing together the learning and efforts of a lot of experts—and we are all experts when it comes to our own places (or we should be)—brings power to the community endeavor. At the same time, even though each person applies his or her individual expertise and interests to a facet of the project, the broader view must be maintained. "It isn't just about water, clean air, or energy consumption. It's also about economic development," explains Ed Schock. A community that doesn't have jobs and economic development for its citizens is not sustainable. As both working committees and individuals bring biases and interests to the table, there must be strategies in place to keep the broader interest of the community at heart. And progress has to be celebrated. When you look around Elgin, there are all kinds of recycling programs. The community needs to recognize the great things going on locally.

For all the successes, at strategic junctures in the process city leaders and citizens have to say, "Wait a minute, is that really what we want to do?" Groups need to reflect periodically in order to learn from successes and challenges. "Elgin is a leader, period," Ed Schock says. "Building a sustainable community is not something that is being done by a lot of communities this size, so what

we're doing is absolutely amazing. When the economy comes back, and it will," he says, "we'll be ready with new programs and policies that are going to help guide us where we want to go." The city can use the lulls in productivity as an opportunity. Put up signs for things that will be built in the future. Get people excited about what's to come, but don't overextend resources. Manage the growth by not trying to do too much all at once. In Elgin the goal is to implement about 15 percent of the Sustainability Master Plan annually, leaving room for changes along the way as conversations change and the community learns from what has already been put in place. Schock calls this process the "weaknesses, opportunities, and regrets protocol."

Elgin has struggled with its economic downturn caused by the long-term unsustainability of having only one big business. Nature does not tolerate monocultures for long. Diversity and a system of organisms are needed for sustainability in nature, just as both diversity and connectedness are needed for a vibrant economy. The vibrant economies of the future, as in Elgin, will be filled with smart, innovative people using local resources; wasting few materials, energy, or by-products; stamping their products with a positive and proud identity; and confidently creating places where people want to work and live. Elgin's goal is to use the concepts of sustainability to be more economically viable. Innovation requires education. In Elgin, community educators and their students are becoming the ambassadors of sustainability. They are using classroom projects and community outreach surveys to open conversations with neighbors that motivate change. After all, people change because it's good for them, not only because it is the right thing to do.

A sustainable community also has a sustainable built environment. The most effective way to have a healthy community is to build proactively. From the outset of design, set priorities for the interaction between buildings and the environment, for the impact of the buildings on the environment, and the interaction of people with the buildings and their environment over time.

Yeng, a student at Elgin High School whom we met earlier, summarizes the effect of Elgin's efforts best: "I have hope for the future. I believe that we've opened people's eyes and have inspired people, helped them realize that there is a problem—a serious problem. People who reflect on our city's progress will pay it forward and inspire others. I have faith. I believe we're heading in a good direction."

And former mayor Ed Schock reminds us that "this is just the beginning and we have a long way to go."

Chapter 8

Ending Car Captivity

Boulder, Colorado

We know how to fix the obesity and inactivity epidemics. People are 38 percent more likely to exercise when they live within one mile of a park. We all know that if we eat smarter—smaller portions and greater amounts of fruits and vegetables—and exercise more, we will not be as fat and unhealthy.

Knowing what to do is easy, but *changing* what we do is hard—really hard—and desperately important. When we want to change our eating behavior, it is relatively easy to make changes in our personal environment: for example, removing the candy basket from the desk or eliminating cookies in the house. But for a whole population to change, small modifications are not enough. We require public will, not always easily obtained, to alter our social and physical environment.

Today, much of the built environment is rigged against us. In many places this environment promotes unhealthy eating and discourages exercise. Our schools are organized to keep students at a desk most of the day,

with only brief periods reserved for physical activity. Most of the nonacademic time is given to eating. And children are transported to schools, either because the schools have been built too far away for children to walk or bicycle there or because parents fear the neighborhood is not safe enough for their children to walk to school alone. The daily drive-and-sit that children endure *trains* them to be personally immobile, to sit and watch. If we tried to set up a system to make Americans fat and unfit, we could not do much better than our current American way of life. Sometimes I wonder whether people have dogs simply to remind them to take regular walks. They are the lucky ones.

So, let us look at the leanest city in the leanest state in the nation, and ask what this city is doing differently to help its residents get and stay fit. When I think of Boulder, Colorado, I think of the Flatirons, the uplift of mountains right next to town. From any place in the city you can bicycle twenty minutes or less and then be off on a hike through the foothills. The Flatirons hike is astonishing. The mountains, views, and waterways attract people who love the outdoors. Our love of using our bodies as nature intended affects our choices about where to live. This city's form and layout, and which of its businesses succeed and thrive, reflect what the citizens of Boulder have identified as important to them, namely exercise and health. In turn this reinforces a culture that attracts more people who want to live in a community that values fitness. (Plate 14 is Boulder's Pearl St. Mall, a street for pedestrians.)

SYMPTOMS

When we think of Colorado, most of us think of outdoor sports, clean air, and beautiful scenery. Colorado seems to be the perfect place to be physically active through bicycling, hiking, and rock climbing in the summer and skiing in the winter. Colorado has the lowest rate of overweight and obesity in our nation; nevertheless, according to the Colorado's Children's Campaign, nearly 53 percent of adults in Colorado are considered overweight or obese, and obesity is costing Colorado almost $900 million in health care costs annually.[1]

Obesity and inactivity are affecting not only the health of individuals but also the health of the state, and the cascading impacts are causing a statewide economic crisis. Obesity-related deaths will soon be the leading cause of preventable deaths, killing more people annually than smoking. I find this astounding, having grown up in medicine, surrounded by patients and family members who were the casualties of smoking.

Is this a genetic issue? Although genetics play a part in people's weight, these factors do not change quickly in a population; changes in obesity rates are due much more to changes in behavior than in our genes. Centers for Disease Control and Prevention (CDC) studies report not only that obesity in children has doubled since 1976 but also that children and adolescents who are overweight or obese are 80 percent more likely than other individuals to be severely obese as adults.[2] As of 2007, 13.1 percent of children ages two to fourteen were overweight and another 12.8 were considered at risk for becoming overweight.[3] The habits instilled in the young carry through to adulthood. In these same CDC studies, 17 percent of Colorado adults reported no leisure time physical activity and minimal eating of fruits of vegetables.[4] Twenty-seven percent of children report watching three or more hours of television a day.[5] Weight gain is also associated with the use of drugs such as steroids and antidepressants,[6] so in a community that battles depression due to weather or where many people are unhappy due to economic or personal issues, or among individuals using steroids, there are additional risk factors involved.

Is this an economic issue? When these rates are broken down by demographics, they are two to three times higher for minority children compared to white children and also higher for low-income children compared to higher-income children. What we also know is that junk food is more prevalent in low-income neighborhoods, and it is often hard to find fresh produce at stores that families can walk to. Is this an ethnic issue? According to CDC research, low-income African American, Native American, and Hispanic children are almost *three times* as likely as their white peers to be overweight.[7] Before assimilation to American culture, these groups were not more likely than others to be obese.

The state of Colorado began the Colorado Physical Activity and Nutrition Program (COPAN), in partnership with nonprofits, government agencies, and health care providers, to mobilize these partners, collect and analyze local data, and create plans to address the causes and effects of obesity and overweight.

Colorado is the leanest state in the nation, and according to Chris Jones, alternative transportation planner for Boulder, his is the leanest city in the leanest state. He recognizes that Boulder has a very active population and that this is partly due to the infrastructure that encourages people to be more active, to walk more, to bike more, or just to be out to enjoy the great scenery and facilities that have been created for the community; but this infrastructure did not just appear. It has a long history of conscious choices behind it.

Boulder citizens have always been active in creating their community. Frederick Law Olmsted Jr., the son of the father of landscape architecture, was invited to Boulder in the early 1920s and encouraged the citizens there to preserve their local creeks and waterways. By 1952, with the opening of the Denver Turnpike and the post–World War II boom, the community started to grow exponentially. Between 1950 and 1970, the population of Boulder grew from just over 10,000 to nearly 70,000.

The quick growth of the community and the popularity of cars began to raise concerns in the late 1960s. In the early 1970s, a poster showing the beautiful Rocky Mountains disappearing behind brown clouds of air pollution reinforced people's sense that a change in thinking about the city's future was needed. Boulder was losing what had attracted people to the community in the first place—the exquisite vistas and crisp, clean mountain air.

Boulder's response was to become the first city in the country to enact an open space ordinance that taxed city residents so that the city government could purchase thousands of acres of land around the city that are never to be developed. That act allowed the creation of an infrastructure of trails for people to enjoy the outdoors. This encouraged people to have an active life without needing a gym—their workout was a natural part of their day-to-day activity. Boulder also chose compact development through its open space tax initiative and used the high density to layer mass transit and bicycling and pedestrian pathways. The city government started this effort from a strong philosophical base, but intentional, and often difficult, decisions created the transportation choices that are currently available.

Boulder is particularly interesting because it is a successful example of policy and public participation leading to incremental change. Many of the city's highlights, such as the Pearl Street Mall (Plate 14), developed gradually over decades. Coupled with the ongoing efforts of the transportation committee, city council, and county commission—which have been staffed by individuals knowledgeable on transportation topics—this gradual process has led to better planning and effective town-grown work.

Boulder is a university town, which helps to shelter it somewhat from economic turbulence and infuses a youthfulness that not all cities have. Because of the city's location and compact development, real estate is expensive and only the relatively affluent are able to live in the city. Others who work in the city or provide services for it must commute in, and the city must continually reevaluate whether it wants to encourage people to commute by car through new highways or to use other means to get to and from work. Of course, when people do not live near where they work there are downsides, including the rapidity with which individuals such as hospital workers, firefighters, and police can respond to emergencies.

DIAGNOSIS

We know overweight and obesity are preventable (Figure 8.1). The 2001 *Surgeon General's Call to Action to Prevent and Decrease Overweight and Obesity* outlined these steps to prevent and decrease obesity and overweight.[8]

- *At home*, reduce the amount of time spent in watching television or in other sedentary behaviors, such as working on the computer. As noted earlier, 27 percent of youths surveyed in Colorado spend more than three hours a day in front of the television. Building physical activity into regular routines makes fitness a part of the day rather than an event.
- *At school*, children should eat breakfast and lunch that meets nutritional standards. Prepared meals should provide options that are low in fat, calories, and added sugars. Fewer than one-third of Colorado children report eating enough fruits and vegetables. Most elementary schools struggle to provide the required daily physical education, including cardiovascular exercise.
- *At work*, find ways to embed physical activity at work sites. Take the stairs at work or ride a bike to work.
- *Within the community*, encourage the food industry to provide reasonable food and beverage portion sizes and to inform the public of the nutritional value of what they are eating, and encourage food outlets to increase the availability of low-calorie, nutritious items. Create opportunities for people to be physically active.

Most of our grandmothers have already told us this. The difficulty comes in acting on this advice. So what does Boulder have that contributes to the solution, and what is the role of culture in Boulder's better behavior?

Figure 8.1 Walking is one way to reduce child and adult obesity.

Source: Graphic by Scott Izen for the Media Policy Center.

How Transport Choices Shape Cities

Kevin Krizek is professor of planning and design at the University of Colorado and director of the Active Communities / Transportation Research Group, a collection of graduate students and faculty who look at how people make residential location decisions and the travel implications of those decisions. "It is all about trying to further the research that promotes active communities as well as active transportation," he explains.

There are several complex issues to examine, but for all families and residents, safety is the primary issue. Convenience is another big element, and there are economic considerations, but to the extent that planning can increase the quality of life and make doing the right thing easy to do, the more we see better life choices taking hold. The people of Boulder have a strong interest in living more environmentally sustainable lives. This is a movement that is beginning to grab hold among people of all ages. It is feeding all kinds of other life choice decisions, and where it is present, it is something communities can use to help people live healthier lives. When leaders and residents are looking at improving a community, they have to look at what they have to work with and the factors that are important to those who live there.

Boulder now has an outstanding network of bicycle facilities, with corridors and conduits through which people can move around the area. When visitors ride the bike paths around Boulder, not only are they impressed by the numbers of people who use these facilities every day, but they also see the smiles (Plate 15). There are almost 200 miles of bike paths and lanes for citizens to ride on within the city of Boulder. Routes designed for bicycles may be characterized either as *on-street* or *off-street*. An on-street bike lane is a roadway lane that is restricted to bicycle traffic. An off-street path is completely separate from motorized vehicle traffic and is intended to be shared by bicyclists and pedestrians (Figure 8.2).

The top three cycling-friendly cities in America, by city size, are Portland, Oregon, for a relatively large city; Boulder, Colorado, for a midsized city; and Davis, California, a quintessential university town, for a small city. What is particularly interesting about these three cities is that their facilities

Figure 8.2 Cyclists in Boulder use a combination of on- and off-street routes.
Source: Photograph from the Media Policy Center.

for bicyclists vary dramatically. Davis offers a plethora of off-street bicycle routes. Portland supplies an abundance of on-street facilities. Boulder provides a blend of these two structures, and in that city researchers are seeing anywhere from 18 to 30 percent of Boulder's adult population bicycling on a good weather day.

Ever since the late 1970s, Boulder has realized the potential of the recreation amenities it has, and so the city started with the streams and tributaries coming down from the Rocky Mountains and developed off-street bicycle facilities primarily in the east-west corridors along these vistas. Then, realizing the limitations of east-west transport, it developed a number of primary corridors running north-south. Those were primarily on-street, and that was what began the remarkable collection of both off- and on-street facilities. (Plate 15 shows an area with both on-street and off-street lanes.)

Today, along some primary corridors, Boulder has established both types of bikes routes. For example, an on-street bike lane may have an off-street path paralleling it closely or just one or two blocks over. Each caters to a different type of user. Families are more likely to prefer the protected routes whereas an athlete rushing to class at the University often prefers the on-street route. Communities have different roads for different car needs, why not for bicycles? There is no one ideal recipe for a city's active transportation infrastructure; it must be organic and adapted to the place, weather, and culture.

Cities created for people using automobiles look different from cities created for people using mass transit.

Kansas City and Salt Lake City are streetcar communities. You know this because when the streetcars were put in place, shops tended to be built where the streetcar stopped. You can see the pattern in how the city evolved. Streets vary according to what goes on around them. Where there is a streetcar, there is a different kind of store—a different kind of storefront.

When our cities are designed for cars, what we get are wide streets that maximize the flow of traffic. I assert that many state departments of transportation do not have a sense of how to make a street that serves a community. They know how to move traffic, and move it faster. They will ruin Main Street by putting in recessed parking (which looks like a notch has been taken out of the sidewalk to create a buffer zone between the flow of traffic and the parked cars. The idea is to keep cars moving quickly down the street. In one city in Wyoming a downtown sidewalk was narrowed to just three or four feet in order to fit in a turning lane for vehicles. This doesn't make for a great downtown, but it does move traffic along.

Planners make roads wider to expand the number of people who can access a place. This encourages shopping malls and big-box stores that dominate an area, eventually destroying local businesses. Yet, as people turn back to wanting more community, some cities are responding with a shift back to a Main Street intended for everyone, not just drivers, and they are examining how people move now and might be moved in the near future. There are encouraging signs out there, but you've got to look.

A Case for Bicycles

Compared to people in all other developed countries, Americans spend the highest percentage of their income on transportation—more than 18 percent of all the money they earn, and that percentage is even worse for the poor.[9] What if a family were to go from two cars to one? This is something that could be done in a bicycle-friendly town. By eliminating one car, the family could save over $6,000 a year, and that's like being handed a check for that amount free of taxes. Having $500 a month to put toward a mortgage instead of a car means the family could afford a better home; they could build their equity, and get tax benefits. The question then becomes how to preserve our income from cars.

National data suggest that the average bike ride in the United States covers from two and a half to three miles. In Boulder, almost every destination within the city limits is within five miles of another destination, which is one of the factors that makes the city bicycling friendly. People can ride north-south or east-west corridors without having to cross traffic lights.

I expect that if you were to ask a hundred people what the biggest factor is that prevents them from cycling, the majority would say that the car drivers are not careful enough near bicycles. The traffic intersection is ultimately one of the most troublesome points for cyclists. Creating bike paths that are separate from streets helps to remove the interaction with cars. My own students who bicycle to UCLA report that in addition to concerns about being hit by cars, they worry very much about being *doored*—running into a suddenly opened door as a driver gets out of the car into the bike lane on the left side of the car. Although the jury may still be out as to whether these separate pathways are indeed safer, they are perceived as being safer by a majority of people—by families, young children, older people, and inexperienced cyclists. A large number of women tend to see them as safer.[10] According to Marni Ratzel, bicycle and pedestrian planner for Boulder, safer routes are more inviting to bicyclists, and also as bicyclist numbers increase, drivers become more aware and more careful.

This is a good place to mention a campaign by my physician friend and exercise enthusiast, Andrew "Andy" Dannenberg. Andy worked for three years in a sixth-floor office and prided himself on only twice taking an elevator: on moving day when he had lots of boxes and on the day after he had knee surgery. He worked in Atlanta and almost always commuted to work by bicycle. He has completed 400-mile bicycle events in many places; one of his favorites is in the Colorado Rockies. He has many passions, but one is the importance of correct placement of *rumble strips*—the line of bumpy transverse grooves found on the outer portion of a roadway. These grooves are there, of course, to alert a driver who is drifting out of a lane. For a long time, however, many state highway departments have been placing rumble strips in the *middle* of bike lanes, instead of between the bike lane and the roadway. Could there be a more effective way to discourage bicyclists?

There are seventy-four bicycle underpasses in Boulder. For a city of twenty-five square miles, having seventy-four underpasses is astonishing. Each city has its own peculiar frame to deal with, and there are different schools of thought on siting bike paths for safety. One school recommends shoehorning bicycle pathways anywhere, even putting miles of bike paths next to a road, but others suggest separating bicyclists from the road to the greatest extent possible. Unfortunately, a lot of transportation options are restricted, but there is an active discussion about what types of facilities are safer. To the extent that communities can minimize road crossings or address crossings to make them safer, they get closer to the ultimate goal of increasing walking and biking.

Marni Ratzel, Bicycle and Pedestrian Planner for Boulder, on the City's Counter-Flow Bike Lane

This is the 13th Street counterflow bike lane [Figure 8.3]. It's about one-quarter mile long and it's in our downtown. It's the solution that we found that works well to keep our bike network seamless. This bike lane connects on-street bike lanes to a multiuse path known as the Boulder Creek Path [Figure 8.4]. It was the way that we could make the connection between a very busy business area downtown and a retail center.

An on-street bike lane is a lane within the road, specific for bicyclists, and there are several of them on the roadway. An off-street path, what in Boulder is called a *multiuse path*, is completely separate from vehicular traffic and is shared by bicyclists and pedestrians.

The other thing that I'd like to mention is *counterflow*. That is when the bicyclists travel in the opposite direction of vehicular traffic so it is a one-way street for motor vehicles. Bicyclists share the road space with motor vehicles in the northbound direction and bicyclists have a lane all to themselves in the southbound direction that is physically separated from the cars. Bicyclists can see the cars headed toward them, rather than cars coming up from behind, which adds another sense of safety and comfort while riding. This is an idea that has become common in progressive areas of Europe, yet Boulder has taken the lead in bringing this idea to North America.

Marni receives calls from communities around the country about how this works. I can say it's actually a very pleasant ride experience.

Figure 8.3   Marni Ratzel at the 13th Street bike lane.
Source: Photograph from the Media Policy Center.

Figure 8.4   The Boulder Creek Path is a major bicycling artery.
Source: Photograph from the Media Policy Center.

CURE

Success begins with getting the right people involved. Tracey Winfrey of the Boulder Public Works Department explains that Boulder started focusing on transportation twenty years ago when the city council, the planning board, and active members of the community all asked, in effect, "How do we want the transportation system to fit into our community?" This was in response to a threat of major roadway expansion through the center of town (which led to an interest in underpasses). The result was a Transportation Master Plan that said, as Tracey Winfrey puts it, "We want to provide not just for cars getting around the community, but for bicyclists and pedestrians. We want public transportation and we want to be very intentional about each of these facets working together." So for the first time, Boulder created a plan that put together people and the resources needed to create the options that were envisioned.

Having a strong policy foundation is very important. That and intentionally putting people and resources to work to create options created a synergy that has been getting results. Boulder has been lucky to be small enough to be nimble—to be innovative. People and groups concerned about transportation have been able to move policy in new directions with political support on an ongoing basis, rather than fighting through a bureaucracy.

I think some communities are just emerging to the point where they have advocacy groups, advisory boards, and city councils that are saying, "Hey, we could be doing this. Let's do it." Then the new approach becomes more institutionalized within the city and county government. Part of change is about getting an approach ingrained until it is a natural facet of the local culture. That is when communities start to see progress. The challenge is that it takes time, and people have little patience.

It's clear that Boulder's transportation initiative is working when people find that the quickest way to move between home and work is by bicycle. Chris Jones explains that he has "chosen to live in Boulder because it means that I can have a normal, quick commute to work where I can get on my bike from the moment I leave my house to the moment I arrive at my office. I don't have to hit one stop sign or one red light. I have a direct bike route that gets me to work faster than a car and that is attractive for me. My commute serves as my workout so it's great that my commute can help keep me in shape."

At the city's Transportation Division, Jones works on alternative transportation planning and a federally funded campaign called Driven to Drive Less. The campaign encourages individuals to give up their car for periods from a month to a year and highlights members of the community who discontinue use of a car. "We show our residents ways that they can get around town without their cars and empower them to leave their cars at home and enjoy the great facilities that we've created for them," says Jones. "All the while they'll save a lot of money and hopefully they'll feel like they've increased their quality of life as a result of their participation in the program." For those who cannot completely give up their cars, Boulder has promoted a Community Transit Network that helps those who do need a ride to connect with others with a similar in-bound commute.

Kevin Krizek describes one of the main themes of his research as trying to understand the different dimensions of cycling and what it takes to get people to cycle and walk more. His group looks at the economic benefits of and preferences for different types of cycling facilities, and the best ways to count and measure cycling facilities and activity. For example, his group "just finished a research project that created a tool to look at the benefits and the costs of bicycle facilities and to weigh them against one another in the event that a community wanted to build an off-street facility and asked how much it would cost and what are the benefits that would likely ensue from the development." Krizek explains that in his research he's "seeing a groundswell of enthusiasm that is related to people no longer being interested in being held captive by their cars. As a result, we're looking to increase modes of transportation and in particular what I'm interested in is cycling. I feel as if we've been working this cause for awhile, but we're just about ready for a turning point to take hold with respect to increased rates of cycling and increased rates of enthusiasm about getting people on bikes." As discussed earlier, Boulder is one of the top cities in America for rates of cycling and its use of both on-street and off-street facilities to meet different riders' needs makes its system exceptional.

First and foremost the development of the bike network aims to provide enjoyment and fulfillment to the individual. Second, it helps take cars off the road to yield environmental or carbon savings from decreased pollution. There are arguments that suggest that it could reduce traffic congestion, but increased rates of cycling are directly related to increased community vitality and increased rates of vibrancy. People are smiling—they are happier and healthier. That is really the benefit you see.

PREVENTION

To increase physical activity, communities need a long-range plan. Kevin Krizek has been watching the development of transportation legislation at the federal level. The Safe, Accountable, Flexible, Efficient Transportation Equity Act—A Legacy for Users (SAFETEA-LU), is the single largest domestic discretionary spending bill in the federal budget and is reauthorized every six years. Funding was authorized in 2005 to provide a total of $25 million in funding to Columbia, Missouri, Minneapolis, Minnesota, Sheboygan, Wisconsin, and Marin County, California, to pilot new nonmotorized transportation facilities.[11] Krizek's Transportation Research Group is responsible for monitoring the efficacy of these investments through pre- and postimplementation surveys to see if an uptick in bicycling appears once the facilities are completed. It is hard to say what a new Congress will enact under the current economic conditions.

Tracey Winfrey and Martha Roskowski from Boulder's Public Works Department suggest that to get an uptick in bicycle ridership, cities must work at the local level to develop a *culture* that supports alternative modes of transportation. Cities must think about how cars, buses, light rail, bicycles, and pedestrians can best travel the city, and then partner with the community to promote the strategies identified. In much of the country communities have begun to make progress in providing travel choices, but there are a lot of reasons why programs have not developed as quickly or as successfully as we need. Some of the challenges are financial, but most have to do with the moving target of congestion. By the time a city puts a new plan in place, the demographics have changed, there are more people than the new structure can support, and there are new priorities that divert attention to other initiatives.

The question is, can a Boulder model be applied to other cities? Boulder has a number of characteristics that tend to make people say of what Boulder does, "Oh well, that's Boulder." Boulder is a university town, it has a very stringent growth boundary, it is wealthy, and it is a small town with an extremely physically active population, including Olympians, marathoners, and world-class triathlon contenders all living right on top of one another.

Notwithstanding, Boulder has lessons to teach. Boulder did not reinvent itself overnight. It takes a generation to make this happen, but there are some lessons that could be applied by other communities:

- The city makes the most of its natural physical features, such as creeks and streams, and aligns them with bike paths. Everyone loves to bike or walk near a creek.
- The city created a series of routes for bicycles between places of interest. Recognizing that not every street can be bicycle friendly, the city focused its efforts on a few paths that act as arteries for moving bicyclists through the city.
- Safe, secure, and convenient bicycle parking is essential. If it takes longer to find parking than to get to work or, worse, if the bicycle is not there in the bike rack when the rider is ready to go home, people will not bicycle.
- Citizens of different ages and with varying ideas should be engaged in the decision-making process. These local voices are needed when planning proposals lack a solid bicycle component.
- Schools must be partners. Children should be able to bicycle safely to school, and adults can then use these routes as well. The more bicyclists there are, the safer each cyclist becomes.

According to Winfrey:

[Boulder gets] a lot of visitors from other cities who come to see what Boulder has done and how they can make their community more like Boulder. A few go away discouraged because they look at what Boulder is today and what their community is today and say, "Oh my God, we could never be like Boulder," and yet I think most of them go away inspired that through a systematic approach, dedication to a long-term plan, thinking through the challenges, and putting the issues on the radar screen, every community can evolve and can improve.

Every community makes decisions about transportation, recreation funding, retail, school locations, housing development, and how the city will grow. If the plan is clear, it is no more expensive or that much more difficult to build a transportation infrastructure that works for everyone, and in the long run it will save money and attract more productive, creative, and energetic people. Boulder has been called "twenty-five square miles surrounded by reality." I say we can all use some of Boulder's reality.

Chapter 9

Ports as Partners in Health

Oakland, California

Money buys a lot of things in this world, including a cleaner and healthier environment. People with higher incomes can afford to pay more for places they want. In contrast, places that are undesirable because of climate or proximity to dangerous, ugly, or annoying features may be the only places where people with less income can afford to live, and this can have a major impact on their health. What is important about Oakland and other major port cities is that even though they support vast amounts of trade and wealth generation, the city inhabitants who live near these ports frequently enjoy few of the benefits a port proposes to provide (Figure 9.1).

In general, port cities reflect the best and worst of an area—they hold the potential of great wealth through economic exchange but they also embody the health risks associated with shipping and manufacturing. Geographical, geological, industrial, social, and commercial barriers divide most cities. Oakland is no exception. Interstate freeways effectively segregate the city

economically, with the financially affluent living in and near and on the Oakland Hills to the east of the highways, and the poorest residents living in the "Flats" to the west, closer to the port and to the major highways.

Special health problems confront the west enclave of Oakland, and their roots are social and environmental. My friend and colleague Dr. Anthony "Tony" Iton has always had a lot to say on such issues. His remarks about his own awakening to environmental and social injustice form a fitting introduction to the theme of this chapter:

My first recognition that the environment has deep and profound influences on people's health was when I arrived in medical school. I first saw East Baltimore when I was being driven around by an upper classman. I was in shock. I didn't understand what I was seeing. There were row houses in which every third or fourth one was boarded up. Some were burned out with blown out windows. Cars were up on blocks and broken glass was all over the place. Mangy dogs

Figure 9.1 The Port of Oakland is one of the busiest gateways for goods in the northern hemisphere. *Source:* Photograph from Media Policy Center.

do you expect? This is the inner city." It hit me—there are people who expect this. Inner cities are walled off war zones where people are left to their own devices.

I started working at medical school, seeing patients from this community. I realized that I was trying to treat social ills with pills and the issues that people were bringing to me had nothing to do with medicine. The issues had to do with a lack of core health resources—access to healthy food, employment, and opportunity. Here I was looking for a pill to treat the manifestation of social deprivation. I immediately saw the environment as being a critical determinant of health outcomes. We had to contend with and recognize that medicine has to look beyond and further upstream—to the root causes of ill health.

Later, Tony Iton served as the chief public health officer of Alameda County, where Oakland is located. We will hear more from him later in this chapter.

SYMPTOMS

Oakland was never *the* destination. It was a crossroads for people and resources beginning with the area's colonization by Spain and Russia; its oaks and redwoods were used to build San Francisco and its fertile flatlands to feed the region. During the Gold Rush it was a crossroads for moving goods and services across the nation. Oakland was the western terminus of the first transcontinental

wandered around, spaced-out looking people sat on their stoops like they had nowhere to go and nothing to live for, and children played amongst this.

I was a twenty-two-year-old naïve Canadian. I asked the driver, "When was there a war here?" because in my mind there was nothing that could compare to this other than images that at the time were coming out of Beirut and other places that seemed similarly blighted and deeply desolate. He turned to me and said, "What railroad in the 1860s; a major stop along the Lincoln Highway, completed in 1916; one of the busiest ports in the world; and the location of automotive factories. In Oakland, people and resources change their mode of transportation from ship to truck to rail to ferry to subway to car, and back again. Oakland is a place people and products move through.

There are a host of "inferiority complex cities." Newark, New Jersey, where I grew up, is one. It is known for being "on the way" to New York City or Philadelphia. It is a tough town, but better than it was thirty years ago. When I think about Newark, I think about the people. They will get in your face. They can be gritty and direct—they are *real*. Oakland reminds me of home—the city does not put on airs or pretend that it is San Francisco.

As industry grew in Oakland, so did the population. Oakland has better weather than San Francisco, and one hundred years ago, West Oakland had magnificent homes and was a good place to live. In fact, following the 1906 earthquake, many San Francisco residents left that city and moved to Oakland.

For work in the automotive plants and in the war industries of the 1940s, Oakland attracted people from the U.S. South and from across the globe, making it one of the most ethnically diverse cities in America. People used ferries and trolley cars to commute between home and work. The nature of public transit kept people in close proximity. Then with the growing popularity of cars, the decline of the trolley system, and the gentrification of major cities, blue-collar workers began moving to *second*

cities. Workers in San Francisco began moving to West Oakland near the port, where they could afford a home, while the wealthy in Oakland purchased the imported goods the port delivered but lived away from the side effects of industry. As African Americans moved into the community, even many working-class whites moved away, and what remained were industry and a community with high rates of poverty and pollution.[1]

Air Pollution

Air pollution in Oakland comes from three main sources: mobile sources, such as cars, trucks, trains, and ships; stationary sources, such as factories and power plants; and area sources, such as fireplaces, lawn mowers, and dry cleaners. About one-third of the diesel air pollution from all these sources comes from the Port of Oakland.[2]

Brian Beveridge, codirector of the West Oakland Environmental Indicators Project, describes West Oakland as a highly industrialized neighborhood. "It has been this way for decades," he says. "It's a reflection of a kind of community planning that happened earlier in this century. The factory was shoulder-to-shoulder with residential housing. This mixed use was a bad deal for the workers and residents due to industrial pollution. In time, regulations cleaned up most of the factories and workplaces, but no one considered the long-term impacts to the neighborhood around those industrial facilities—the air, water, and soil."

As mentioned previously, Oakland is an extremely busy port (Plate 16), off-loading more than 99 percent of the imported containerized goods moving through Northern California. The port's cargo volume makes it the fourth busiest container port in the United States. The combined flow of cargo into the ports of Oakland, San Francisco, and Los Angeles equates to 50 percent of all goods entering the United States.[3] However, the majority of the people who live in Oakland do not even realize there is a port there. It is not the focus of the community. The population living within the triangular area near the port that is surrounded by three highways is small, but it is disproportionately affected by industrial and commercial activity.

Terry Smalley, who manages the Crane Department for the Port of Oakland, where he has been for forty years, points out that "at the port, we have cargo that comes from all over the world and moves into the United States. We have about 1,900 container ships a year transporting about $33 billion worth of cargo." That is a lot of important commerce, but there is a cost associated with that, too, says the former chief public health officer of Alameda County, Tony Iton. "Ships burn bunker fuel, a high-sulfur fuel that generates very noxious diesel particulates. Diesel trains come into the port, pick up the containers, and ship them all around the state. We've got trucks that line up at the port that are burning diesel fuels and are idling much of the time, keeping their engines running." Commerce in Oakland is creating a public health crisis. "The load of soot in the air in West Oakland is three times the county level. The disease rate is overwhelmingly concentrated in these areas in the form of heart disease, asthma, cancer, and other forms of chronic lung disease," Iton says. The soot from the air settles on the land and water. The build-up of particulates in these areas doesn't dissipate and it ends up in the food people eat and the water people drink.

Access to Healthy Food

West Oakland has had three supermarkets come in and fail. Remarkably, the small mom-and-pop stores have done fine, but residents cannot get much fresh produce or meat at these markets. I am told that if they have access to a car, they drive to the big-box store supermarkets for a big provisioning. People going to the local convenience stores are walking and they are buying only what they can carry. They are going to see more pastries and beer than fruits and vegetables. They will purchase what is inexpensive and available.

One way to create changes in food sources in an area is through restrictive zoning. After the 1992 riots in Los Angeles, about half the liquor stores closed and crime dropped dramatically in the area. The Planning Healthy Places team within the advocacy organization Public Health Law & Policy has defined a number of desirable zoning requirements.[4] For example, it is a zoning decision to allow a liquor store or fast-food place to open. Communities can write restrictive zoning to create the kind of commerce they want to encourage.

To encourage healthy behavior, we need to eliminate environmental constraints, but we also need to see that community resource facilities are adequately staffed. Many of our schools have facilities that are underutilized because staffing is not available to keep their doors open. Without after-school activities at schools, what are kids doing in the afternoon? Often they are hanging out by fast-food places or playing video games at home. Without viable options for healthy activity, they are learning a pattern of lethargy.

Unintended Consequences of Poor Land-Use Planning

When we build today, we also build on the past; the history of an area can be an asset or a threat or (commonly) both. As Brian Beveridge says of Oakland:

> We live with pavement and sidewalks, buildings, factories, smoke stacks, buses and diesel trucks. The products of this built environment are the canvas we have to work with in an attempt to improve our city. We have a legacy of old housing and industrial uses, truck parking lots, foundries, and paint manufacturers. The soil is full of toxins. Anytime we want to build something we have to consider how the soil has been used in the past and what we're going to do with it. Cleaning it up is hazardous to the people who live around it. The environment our ancestors built around them has an impact on our quality of life even now.

DIAGNOSIS

When Tony Iton was chief public health officer of Alameda County, he and his staff reviewed nearly 100,000 county death certificates issued between 1990 and 2000 and mapped where people died. They calculated the rates of death and the ages at which people died in different communities (Figure 9.2). Iton comments:

> We found dramatic differences in the most urban areas. We saw life expectancies fifteen years less in urban areas than in the hills of Oakland or in other parts of Alameda County. We recognized that this

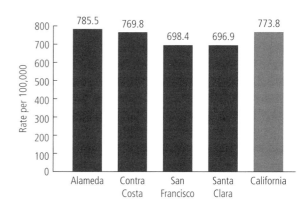

Figure 9.2 Death from all causes: selected counties and California: age-adjusted rates, 1998–2000 average.

Source: Alameda County Public Health Department, *Rates of Death from All Causes* (2003), http://www.acphd.org. Used with permission.

wasn't because the behaviors or the people were different. We noticed that people lacking access to basic resources like fresh air, parks, and public transportation, made a difference. These things have a deep and rooted influence on health outcomes. How we live shapes how long we live. We tell people in our county, give us your address and we'll tell you how long you'll live. We are right most of the time.

According to the data Iton compiled, if you want to live a long life in Alameda County, you want to live up in the hills of Oakland or you want to live and work out in the suburbs, in Pleasanton or Livermore (Figure 9.3). There is more wealth in these areas, and frankly, wealth equals health in this country. *Where* we live and *how* health is distributed shapes the difference in our health outcomes.[5]

We know there are physical and environmental conditions that strongly influence health outcomes. In 1969, the National Environmental Policy Act (NEPA) set a standard of safe healthful environments for all Americans and required environmental impact assessments on federal projects. In 1983, the U.S. General Accounting Office further raised awareness on the issue with the study *Siting of Hazardous Waste Landfills and Their Correlation with Racial and Economic Status of Surrounding Communities*. This shed light on the need to mitigate disproportionate pollution concentrations and involve communities in enforcement of environmental standards. As a result of the increased awareness, in 1994 President Clinton issued Executive Order 12898, which strengthened the assessment of exposure, risk, and impacts of policies on specific populations; however, this Order addressed only federal projects and did not include commercial or industrial emissions. Local environmental justice groups are now encouraging business owners and the general public to apply pressure toward a wider application of these policies.

According to the Pacific Institute's report *Clearing the Air*, "Freight transport in California is powered almost exclusively by diesel engines, many of which are old and dirty. The diesel trucks, trains, ships, and equipment used to move goods around the state emit numerous

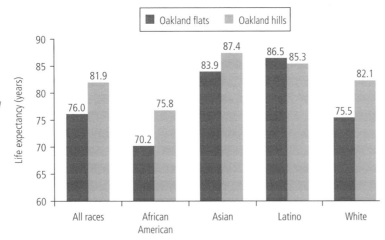

Figure 9.3 Life expectancy at birth: Oakland Flats and Hills, 2001–2005.

Source: Alameda County Public Health Department, *Life and Death from Unnatural Causes: Health and Social Inequity in Alameda County* (Apr. 18, 2008), http://www.acphd.org/healthequity/reports/index.htm. Used with permission.

pollutants. Diesel exhaust can contain an estimated 450 different chemicals, 40 of which are listed by the California Environmental Protection Agency as toxic air contaminants that are dangerous to health even at extremely low levels."[6]

There are six times more diesel particulates emitted per person and over ninety times more diesel particulates per square mile per year in West Oakland than in the State of California as a whole (Figure 9.4). There is seven times more diesel exhaust per person in West Oakland than in Alameda County as a whole. The amount of toxic soot produced by trucks traveling in West Oakland in one day is the same amount produced by 127,677 cars. In 2005, freight transport activity (excluding air cargo) contributed about 30 percent of the total statewide NO_x emissions and a stunning 75 percent of all diesel

particulate matter emissions in the state. Diesel particulate matter is among the most toxic air pollutants. Diesel contains toxic and carcinogenic compounds such as benzene, arsenic, and formaldehyde. These compounds can go deep into the lungs and directly into the bloodstream. Short-term exposures make allergies, asthma, and chronic bronchitis worse. Long-term exposure leads to increased rates of cancer. Some West Oakland residents are exposed to roughly five times more diesel particulates than residents in other parts of Oakland are.[7]

If we know the sources of air pollution, why don't we do something about them? Technological advances and regulations aimed to reduce air pollution are often offset by the increased numbers of commuters on the road, the greater number of miles each commuter travels, and the higher volume of goods being transported in and out of an area. Congestion increases idling time and this in turn decreases the quality of the air. Cars and trucks are burning fuel more cleanly and efficiently, but at the same time, commerce has been increasing. (Plate 17 shows the West Oakland skyline, bounded by the port and crisscrossed by freeways.) The tectonic plates of behavior, economics, and health risks continue to shift, and the net change in air quality is minimal.

Two big public health issues affecting Oakland, air quality and obesity, could both be reduced by pushing a single strategy—increased use of light rail transit (LRT). In September 2010, a new study reported what happened among people in Charlotte, North Carolina, after a light

Figure 9.4 Average diesel particulate emissions per person, by region.

Source: M. Palaniappan, D. Wu, and J. Kohleriter, *Clearing the Air: Reducing Diesel Pollution in West Oakland* (Oakland, Calif.: Pacific Institute and the Coalition for West Oakland Revitalization, Nov. 2003), p. 2. Used with the permission of the Pacific Institute.

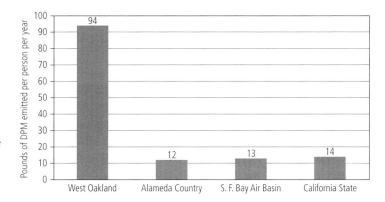

rail system was put in place. People who were using the light rail—and they were not doing this as part of a weight loss or exercise campaign—weighed less and felt better. They had a more positive perception of their community, they walked to and from the LRT, and their body mass index decreased. For a person who is five foot five, the average loss would be 6.45 pounds. Those using the LRT were 81 percent less likely than nonusers to become obese over time.[8] (Plate 18 displays some of Oakland's BART-centered, transit-oriented development.)

Identifying Social Inequity

When I was at the Centers for Disease Control and Prevention (CDC), my department funded a project with Larry Cohen's Prevention Institute to do a set of local community profiles on how the environment could be changed to advance health. By asking questions related to equity we could begin to understand the characteristics of middle-class suburbs, low-income communities, communities of color, inner cities, and rural communities. By examining these different profiles, we could better determine where a community lies on the spectrum of social equity.

It is estimated that the amount of freight moving through the Port of Oakland will increase if and when the economy rebounds. Even though the commerce is economically beneficial for the state and for business, the air quality levels may spell a different outcome for local residents. As an Alameda County Public Health Department report put it, "social inequity causes health inequity."[9] When residents are mapped against the location of known polluters, nonwhites, recent immigrants, and the poor are more likely to live closest to air-polluting facilities, their numbers dropping as the radius from these sources increases. Tony Iton describes what this means for all of us:

> There's an equity issue because the biggest shippers to the Port of Oakland are the big-box stores and they are making profits through a global economy on the backs of the health of some of the least politically, economically and socially powerful people in this country. These people are suffering disproportionate levels of asthma and other chronic diseases [Figure 9.5]. Ultimately, these equity issues involve us all and we all have to come up with solutions that spread both the benefits and the burdens of our way of life in a more equal way.

Building for Economic Integration

One idea is to encourage mixed-housing communities. Where communities have a mix of people of both lower and higher socioeconomic status, the wealthier will demand services and implementation of existing air quality regulations. They will demand restaurants, places to shop, supermarkets, and better policing. All the things that people with less money need too. To keep an

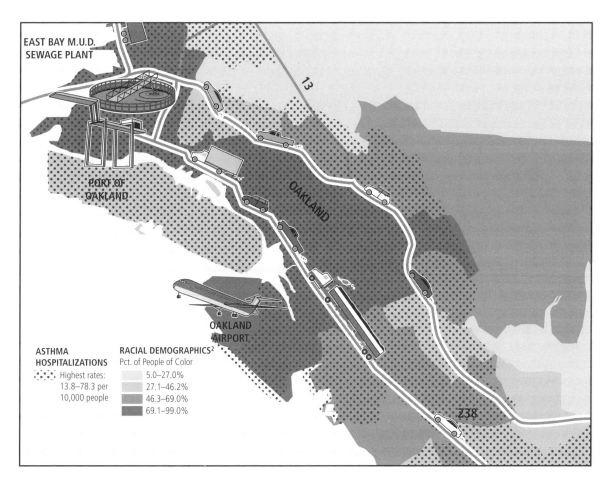

Figure 9.5 Asthma and race in Oakland.

Note: Asthma data are from Community Action to Fight Asthma, California State Coordinating Office, *Asthma Hospitalization Rates (1998–2000 CA OSHPD Data) by Zip Code* (Oakland, Calif.: Community Action to Fight Asthma, 2004); demographic data are from the U.S. Census 2000.

Source: Bay Area Environmental Health Collaborative, http://baehc.org/resources. Used with permission.

area from becoming too exclusive, regulations for low-cost housing can keep a mix in place, as evidenced in Sacramento or Santa Monica in California.

Oakland implemented a campaign to add ten thousand condos to the community. It is great that more middle- and working-class homes were added, but the design seemed a bit cookie-cutter. They were almost all two-bedroom condos, no one-bedrooms and no three- or four-bedroom units. With a variety of housing options, a community sees much more diversity of income in its residents. If it has diversity of income, available services are better. As I tell my students, "everyone knows that rich people need poor people to provide local services, but the reverse is also true in terms of providing a tax base, policing, and other social services."

When we decide we want to do something about air pollution, it takes a while before we can see the benefits of a change. Regulations affect primarily newer equipment and capital improvements and can take several decades to positively influence the health risks posed by pollution, including diesel pollution. Meanwhile, people continue to get sick, miss school or work, or become hospitalized. After the Toxics Release Inventory (TRI) in Oakland began reporting levels of pollutants in 1987, Oakland overall showed a great reduction in total toxic air releases. The power of the TRI lies in the act of disclosure to government, the public, and those working for polluting companies, especially at the executive level, about what and how much toxic material is emitted. This national requirement to report drove down emission levels in most areas. However, when we look just at West Oakland, we see a rapid increase in levels, over thirtyfold, since 1996. This has occurred because even though the large polluters in other areas of Oakland were closed down, those in West Oakland were not. In addition, only the large polluters are regulated by the emissions standards, and they are only a fraction of the sources affecting air quality in West Oakland.

Asthma is epidemic in West Oakland. Children there are seven times more likely to be hospitalized for asthma than the average child in the rest of California. Multiple recent studies have shown that diesel exhaust will not only make asthma worse but may actually *cause* asthma. Children are particularly susceptible to diesel exhaust and other air pollutants because they breathe more air per pound of body weight, they take in more air per minute (they breathe faster), they exercise more, and they tend to breathe through their mouths, so the protective filters of the nasopharynx do not have the chance to screen out particulates.

Many West Oakland schools are located near and downwind of busy roadways. Children at these schools have higher rates of asthma symptoms than other children. Dr. Rob McConnell, professor of preventive medicine and deputy director of the Children's Environmental Health Center, did a study of children in Los Angeles[10] that tells us the location of these Oakland schools is a serious concern. He studied student athletes at six Los Angeles schools in high-pollution areas and six in low-pollution areas. These students were active and they ate well. After five years, the student athletes in the high-pollution areas had rates of asthma that were more than three times the rates among the student athletes in less polluted areas. When we allow this much air pollution, we are punishing kids for doing what they should be doing—living physically active lives.

Although public schools could not be built in these locations today, there are *ten elementary schools* adjacent to busy roadways in West Oakland. Children are breathing in car and truck exhaust all day. These schools have higher percentages of asthma cases than schools in surrounding areas. Increased asthma symptoms lead to an increase in absenteeism.[11] Teachers cannot teach a student who is not there, so student achievement suffers. Intervention is costly. Children fail to meet their full potential, and not only does the child lose, all of society loses what that child could have accomplished.

So I ask, what would we do if we really loved our children? Where would we start to build a healthy community? Whose responsibility is it to give the children of the portside poor, the dockworkers, and the day laborers clean air to breathe?

Social Injustice, Racial Bias, and the Built Environment

Contra Costa County is northeast of Oakland. On average it is a relatively affluent community, but if we look closer we will see wide disparities in income across the community, and these income disparities result in different health outcomes. Wendel Brunner, a physician with a PhD degree in physics and the director of public health for Contra Costa Health Services, comments:

> I think the most fundamental inequity you can have is the fact that residents of low-income areas can expect to die measurably sooner than those in the affluent communities. African American children are hospitalized for asthma at about four times the rate of Euro-American children. There are equally shocking disparities in outcomes of diabetes, cardiovascular disease, and the whole host of generally chronic diseases that are responsible for terminating our lives and shortening our enjoyable, youthful years. Ethnicity in this country is heavily associated with income and education. We've noticed that among African Americans in particular, there are much higher rates of diabetes and cardiovascular disease. Hispanics have higher rates of teenage pregnancy. Euro-Americans have higher rates of breast cancer. Disparities can cross ethnicity, but the common theme that underlies all of this is income and education.

Income affects the kind of physical environment that people live in. The environment reflects differing opportunities for physical activity, access to healthy foods, and diversity in eating choices. Exercise and eating habits impact long-term health outcomes. Income also affects access to health care. Low-income people tend to be among the uninsured or the poorly insured. They delay preventive health services and often do not get the same quality of treatment that a higher-income or Euro-American resident might get. Where people use fewer health services, there are fewer medical offices and other facilities, so when services are needed transportation becomes an issue.

Asking these questions can reveal the impacts a built environment is having on a particular group of people:

- Does this group of people have access to adequate shelter at a price that leaves enough money for good food, a proper education, and a variety of other activities? In the built environment, the primary issue is access to affordable housing of reasonable quality.
- Does this group of people have access to fresh fruits and vegetables, variety in their diet, and access to good quality and relatively inexpensive foods?
- Does this group of people have nearby access to work, play, health care, and food resources?
- Does this group of people have opportunities for physical activity and physical exercise?
- Does this group of people have a safe route to school for their children?

Wendel Brunner deals with challenges in Contra Costa County similar to those that Oakland residents battle, pointing out that both "Richmond [in Contra Costa County] and Oakland have a lot of difficulties with air pollution. Diesel trucks and trains move goods from

the port in Oakland through Richmond, and out into the box stores in the Midwest." What commerce does not understand is the relationship between transportation pathways and long-term education trends. Brunner lays out that relationship: "There's probably nothing more important for a person's health than the education that he or she gets. Diesel particulates cause and exacerbate asthma. Asthma is the leading cause of children losing days at school. Education is probably the key determinant of a person's health in the long run." So, air pollution associated with the built environment and poor transportation choices leads to asthma, keeps kids out of school, and can have a major impact on their health, disrupting their education.

Ironically, in Oakland and all across this country some of the newest Hispanic immigrants are some of the healthiest people. They come from communities and countries where the diet is much more healthy than ours and where people are used to physical activity. Once immigrants come into a U.S. community, however, and they acclimate to our environment and our lifestyles, they develop the same health problems and diseases as the residents who've been a part of the community for a long time. The rates of obesity, diabetes, hypertension, high blood pressure, and cardiovascular disease all begin to increase. By the second generation, Hispanics in this country have the same poor health outcomes as the rest of us. So the issues being discussed here cannot be put down to ethnicity; they are a result of our built and social environment.

Food Availability and the Built Environment

Dana Harvey, executive director of Oakland's Mandela Marketplace, explains the food crisis in her community:

> West Oakland has about 25,000 residents. There are no grocery stores in the community; however residents face a proliferation of convenience stores and fast-food establishments. The convenience stores feature alcohol and junk food. Fast food features high fat, salt, and sugar. It is a real challenge having access to healthful foods. . . .

> This is a low-income community primarily because of redlining and urban "renewal" that happened some years ago that drove a lot of the community-based businesses out. We opened the Mandela Marketplace, a nonprofit focused on incubating community leadership through economic empowerment and public health, to provide healthful local, fresh foods to a neighborhood that really has no access to these affordable types of foods. By investing in the residents, we are rebuilding and improving the quality of life in the neighborhood.

It's critical to find the right expertise to move to the next step. Mandela Marketplace is a worker-owned cooperative, so everyone who works in the store owns the store. It's almost a training ground for moving to a bigger store and more responsibility. As the project grows, Dana Harvey's team is focused on building jobs through empowerment, supporting entrepreneurship, and building a sustainable model that reflects the interest and culture of the community.

Forty percent of the store's produce comes from local farmers who have developed a relationship with Harvey over the years. Store workers pick up fresh produce from the farmers on a regular basis so the store is able to get pesticide-free and local produce. Mandela Marketplace works with marginalized, resource-limited minority farmers to make a bridge between the farmers and the community. According to Harvey, "[When we can] say, 'Okay, Mr. Veng, we can buy X hundreds of dollars from you today, so you stay on your farm and we'll come pick up from you,' it's a real benefit both to the farmer and our community."

In order for people to have access to the goods and services they need to be healthy, communities and public health agencies may have to teach them what this means and how to do it. People often have the passion, talent, and skill to own their own businesses, but they haven't had the opportunity. Mandela Marketplace was designed to start providing that opportunity, but this isn't going to change the community overnight.

"You know, I like to think that we are making progress," Tony Iton explains:

> I like to think that the message is getting out, but you only have to go down to the box store, stand outside, and watch what people come out with. As you watch you see inordinate rates of obesity, essentially the result of inordinate consumption of high fructose corn syrup

and heavily processed foods. We have a long way to go to get the message to people about what healthy is and to create more barriers between some of these very unhealthy foods and the populations who purchase it because it's inexpensive and has a high caloric value. What people are buying is actually a pretty good short-term bargain. However, in the long term, it's not such a great bargain. People have to understand the trade-offs they're making due to their lack of resources. There has been some progress, but our practical ability to implement what we know is still in its infancy.

CURE

Increasing Economic Opportunity

Larry Cohen of the Prevention Institute describes Oakland as "a city of enormous challenge but also of enormous opportunity." Oakland is the hub of the Bay Area. There's an artist's depiction of Oakland that I like, showing Oakland as part of a tree of transportation. San Francisco forms the branches and leaves and the southern counties are the tree's roots. This is a great example of how art can change people's perception about a place. When we think of Oakland as the trunk—the central support—it's not a *secondary city* anymore.

West Oakland is a community that has scared some people, and it's a community with a lot of disinvestment.

This is, nevertheless, a time of emerging opportunity for Oakland. Larry Cohen says:

As the economy starts to improve, some of the things [The Prevention Institute has] catalyzed in terms of advancing economic development will be very exciting. Our office is located in a remarkable building. It wasn't quite abandoned but had minimal use before we purchased it. It's an old warehouse built alongside the very end of the Southern Pacific railroad, but now it's a unique, revitalized place. It symbolizes all the best of Oakland—the opportunity, space, creativity, diversity, ethnic and thought diversity. Some of the most important national health and community organizations are based in Oakland. There's a real strong will for Oakland to change.

When I was talking with Tony Iton about Oakland, he told me that what he sees is "economic opportunity," and he explained what he meant this way:

There is no question that we're all connected around the world—the products that come through the Port of Oakland come from every corner of the globe and the products that go out through the port go to every part of the globe. This movement of goods is critical to our way of life. There is absolutely no question about it. We are not going to turn that away, and we shouldn't. But the way that we do it, the impact that has on the people who are actually living in the immediate proximity of these operations, is a reflection of our values.

As Tony and I look at how the United States is changing we have come to realize that this country needs different kinds of people working in public health than it used to. Public health is not just about doctors and nurses anymore. It needs bright young people with laptops, who understand and appreciate the importance of statistical analysis, working closely with other bright people who understand how the built environment contributes to health.

In Alameda County, the public health team has changed dramatically by hiring people with skills in demography, community organizing, and outreach, abilities particularly useful in low-income communities of color, communities that in many cases have more stress than others and less time and fewer resources for organizing themselves in ways that could lead them through a visioning process toward the kind of environments that people want to live in. "We want to see an increase in social, political, and economic power because, ultimately, those are the levers of change in the environment," says Iton. "We need to work in collaboration with people, in organized ways, to create the kind of power that we think leads to change. That's pretty different from what you learn in public health school, but it's fundamentally the root of what we need to do in order to change the equity conditions that ultimately lead to unhealthy outcomes and chronic disease, obesity, diabetes, cancer, and the like."

Pills don't treat air pollution. In Oakland, public health teams are looking at air pollution and trying to understand how it is generated and how it leads to chronic disease. It is not just a matter of finding the right pills but of finding strategies to both reduce the amount of air pollution and medicate its remaining consequences. That means all of us have to think about how our communities are functioning and how design decisions have health consequences.

Marice Ashe, executive director of Public Health Law & Policy, supplies an example: "Oakland is a complicated community. There's been a community coalition to change land-use and truck idling laws to reduce air particulates; however, many of the truckers are low-income immigrants, and they don't have the kind of money needed to upgrade their equipment. The Ditching Dirty Diesel campaign utilizes residents to count, identify, categorize, and monitor trucks through a participatory research project providing data on truck-related activity to be used for enforcement of the California idling law and commercial versus residential zoning restrictions."

Economic empowerment is a cultural shift. It requires people to have dreams, self-esteem, information, and the willingness to work hard. When people see that it can be done, they are more likely to take small, calculated risks and try it. Through city planning initiatives, the types of businesses that Oakland wants to attract can be encouraged.

Another arm of this process is looking at the industries that already exist in the community and finding ways for them to work smarter and cleaner. The port is a perfect example of an area where Oakland can use existing technologies to accomplish the same task more economically while also mitigating environmental impacts. The Port of Los Angeles, for example, has been working on implementing environmental safeguards for some time, including projects in air quality monitoring, alternative maritime power, clean trucks, water quality monitoring, protection of wildlife habitat, aesthetics, and retrofitting existing diesel-powered, on-road trucks. By examining these programs and those at other ports, Oakland can determine the approaches that are best suited to its situation.

Margaret Gordon is an Oakland resident who has gotten involved as a community organizer and who knows the port issues the community is facing. The port administrators, having built better systems of communicating with the residents, now know the issues better as well. Gordon says, "Whereas we couldn't sit down with staff of the different bureaucracies and agencies before, we now go to city hall and bring our issues to the administration and the mayor's office. Our issues are supported through partnerships with businesses and transportation. We're on the same side. We want them to have a successful business because that is our economy, but we have to work together to understand the impacts we have on each other. It's not supposed to be adversarial."

The cranes at the Port of Oakland are now electric powered. Crane Department head Terry Smalley says, "I have seen a major shift. About six years ago, the cranes became electrically fed instead of diesel powered, so that significantly reduced air pollution. The Port of Oakland is making a major push to improve the environment from the perspective of the shipping industry."

Recommendations made by the Alameda County Public Health Department in 2008 include reducing the number of diesel-powered trucks in residential neighborhoods, enforcing the no-idling law near schools, requiring the use of clean technology in new ships and trucks to reduce emissions in existing fleets, and implementing existing state and federal emissions reduction regulations. The trick is how to make these things happen. The solutions originate from a process of making the work at and around the port more transparent. By engaging the community in decision making about locally wanted and unwanted land use and incorporating public health input on air pollution impacts in local land-use planning and development decisions, governmental and nonprofit agencies and community organizations can find new solutions, and monitoring implementation becomes everyone's responsibility, not just the work of compliance monitors.

Reducing the Impact of Violence

One of the pivotal issues in any community is safety. The health and safety of an environment affects people. Violence is a factor in where you live, walk, shop, and work, and even in whether you want to walk, are able to shop, and can get to shops or workplaces or whether you can afford to live in a particular community or not.

"Our airports are safe so our communities should also be safe," says Larry Cohen. "The key in changing violence is not one particular program but a strategy. It's political will—capturing resources."

A lot of work has been done in West Oakland to build community. Communities thrive where strategic investment sparks the growth of new business. During the 1989 Loma Prieta earthquake, the double-deck Cypress Street Viaduct on a major freeway collapsed—it literally pancaked—killing fifty-eight people. Out of this tragedy came major change. The city changed the traffic patterns, and what resulted was Mandela Parkway. In time, new neighborhoods started springing up. Community started happening. Liberated from a freeway that had radiated a message of "stay away," this West Oakland area is now doing much better than it was. Investment in the area has begun, and groups like the Prevention Institute have started to measure encouraging shifts in the pulse of Oakland.

In 2009, on New Year's Eve, a police officer aboard a BART train shot and killed a young African American man, and everyone was concerned about what would happen in the Oakland community. It was clear that the officer had overreacted and he was ordered to trial. Then the trial had to be moved to Los Angeles because the atmosphere seemed so inflamed it was thought the officer could not receive a fair trial from a local jury. This was a test for the community, one that would show whether anything had really changed. A great deal of work went on in Oakland to prevent the situation from getting out

of control, and the Prevention Institute sent out communications to the community, keeping people informed on the progress of the case.

Everyone expected Oakland to break into riots when the verdict came out and the officer was found guilty only of manslaughter, but that is not what happened. There were a lot of positive things in the ways the community reacted to the verdict, even though these responses went largely unreported. The national media covered every broken window, but when the locals reported what was really going on, it turned out that the vandals responsible weren't Oakland residents. There were about eighty arrests, and 75 percent of them were of people from outside Oakland; just a handful were of local people. "In fact," says Rachel Cushing, a teaching assistant doing an internship at the Prevention Institute, "when people saw others from outside Oakland looting, a number of youth from Oakland stood up and went out on the street and said, 'Go home, this is our community. Don't ruin our community.' That was a response from the Oakland youth themselves, and it makes us really proud."

PREVENTION

As Wendel Brunner says,

Public health is all around us. It is what we do as citizens and neighbors to improve the health of our community. So we have local public health departments that are focused on public health, but public health is also what the Parks Department does when it builds parks where we can relax and get physical activity and physical exercise. When the parks are suitable for old people and little children and everyone in between, they work. Public health is what the transportation agencies do when they make public transportation that will help clean up our air and get us safely and comfortably from one place to another, to reduce the stress in our lives and give us time to spend with our families and our children. Public health is what the schools do when they educate our children. We know that education is the most important thing for long-term health.

I think cities like Detroit, Newark, and Oakland have much to teach us. They have had to absorb great population shifts and have been producers and transporters of immense quantities of manufactured goods, but often they have been cheated of the good that comes from commerce—and I am not speaking about so-called goods. Goods for elsewhere should not require ill health among those who make and transport them. Those who have done so much for this country deserve a fairer deal, especially when it comes to their health and their children's health.

Larry Cohen has been an influential contributor to the emphasis on preventive care in President Obama's national health insurance plan. When Cohen talks about the money spent on dealing with challenges such as air pollution, he looks at the pipeline of funding and the end to which most of that money is funneled. His idea is that we need to be investing in prevention. Currently, we take care of prevention last. We spend most of our medical dollars on the final stages of the disease process, on *tertiary treatment*. It is good that we care for the critically ill, but it's not where our main focus should be. We need to be helping people to avoid becoming ill in the first place.

When we spend some money on secondary treatments, like immunizations and antibiotics, we reduce the number of people who actually contract diseases. Investment here does reduce costs later. However, if we want to do the best job possible of managing costs and minimizing suffering, the long-term effects of disease, and premature death, we must do primary prevention. We must develop effective protections against certain viruses and create environments where, for example, it is far less easy for children to overdose on fats and sugars, thus preventing them from becoming obese or developing diabetes. Putting money into prevention is the most effective way to reduce health care costs. This is the fundamental tenet of public health.

To think of it another way, look at an immense organization like the U.S. Department of Defense. All of its vast budget could easily be consumed in fighting wars, the ultimate purpose of the agency, but the Defense Department does not cut off its research or its money for intelligence, for West Point, or the Naval Academy—the long-term stuff. There is no survival for any organization, for any company, and especially for our health care system if all the money goes to the dealing with results of problems that should never have materialized. We have to protect research and development, future leaders, and population-focused interventions. So what Larry Cohen and people like him have been saying is take the same approach with health care as we do with defense. President Obama listened, and his plan makes much disease prevention essentially free.

The public health profession has been struggling to understand how to motivate people to change. We know how to treat individuals, but the treatment of a community must be communicated differently. A doctor talking to a patient can suggest that he or she exercise more, eat less of this or that, reduce his or her stress level, and so forth. We cannot say that to a community. People have to deal with the environment in which they live, so we disseminate information, but direct messaging is rarely effective. We have to change the way we ask questions to raise consciousness.

Everyone involved in the effort to help people to stop smoking has learned a lot from it. Just telling people not to smoke does not really work. Strategies had to be developed to actually reduce the amount of smoking. California has been a great leader in this area. Rates of smoking have declined dramatically in California, outpacing the decline nationally, and rates of smoking-related disease have declined in conjunction with the reduction in smoking (Figures 9.6 and 9.7). Many prevention efforts to come will find us looking at scenarios similar in complexity to the problem of smoking and quite different from

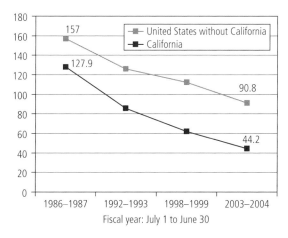

Figure 9.6 Adult cigarette consumption: California versus the rest of the United States, 1984–2004.

Source: Data for packs sold from the California State Board of Equalization and for population from the California Department of Finance.

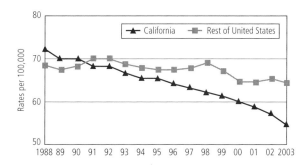

Figure 9.7 Lung cancer incidence: California versus the rest of the United States, 1988–2003.

Source: California Cancer Registry, California Department of Health Services.

those we faced when health care was focused largely on acute illness, and they will require innovative and multi-faceted efforts from us and our communities.

I think young people are the key to changing the health of a community. I think the earlier we introduce kids to how their food is grown and introduce them to nature, the better the understanding they will have of how they fit into the environment. Conversely, the kind of dislocation from the environment that we see in many individuals and communities currently has consequences in terms of lack of stewardship and in poor health.

Wendel Brunner describes reaching a future when "a healthy citizenry is an involved citizenry. It's about getting the residents in the communities to become active in improving their own health." He urges people to talk together about the kinds of changes they would like to see, and to understand what would work to improve their health. All of us need to talk about the kinds of food that should be available and how we can improve the health-giving aspects of our food and still meet the expectations and ethnic traditions that food represents. We need to talk about how neighborhood parks are designed and how they could better serve the community. We need to talk about the schools and how parents, teachers, and the school system can work together to improve the schools, to improve the outcomes for the children. We need to talk about transportation systems and what kind of public transportation we would like to see and need to build. We need to work with our policymakers so that they can understand the directions we want to go and the kinds of investments we need to make in our future, and we all need to reach an understanding that investments in the future and in our well-being cost money. "This work will require us to pay taxes," Brunner says, "but those taxes need to be well spent so that we can enjoy the benefits of long, happy, and healthy lives." One specific example of a useful local investment is streetlights, because as Chris Kochtitzky, the CDC colleague with whom I wrote my first article on the built environment, points out, "Streetlights really do get at obesity, because people are less afraid to walk around their community for fear of getting mugged or hit by a car."

In West Oakland and other communities across the country, we have made it impossible for children to walk or bicycle to school, closed school playgrounds after school due to concerns about liability, and made the parks too difficult to get to or too dangerous to play in. In retrospect, we have committed a whole host of ill advised or poorly executed actions that make it even harder for us to combat the epidemic of obesity that we are seeing now.

Chapter 10

The City That Won't Give Up

Detroit, Michigan

Detroit is the leading edge of the Rust Belt. The city that once made millions of people prosperous building cars for the world is now in economic depression. As the auto industry moved south and offshore, the good union-wage jobs that supported hundreds of thousands of families disappeared. Some neighborhoods in Detroit are empty, with houses boarded up or burned down. In 2010, you could buy a house in some parts of Detroit for less than $100—and that price was likely more than the house was worth.

As Detroit goes, so goes much of the industrial heartland of America. A city built on working-class wages and based on the automobile has withered as times changed. But there is more to the story of Detroit than faded glory and dismal prospects. People are working to revitalize the city and rebuild it along new lines. The same "go to work" attitude that built the place they called Motor City is now being put to use in a new direction. Citizens in Detroit are remaking their built environment to reflect the values of the new century and new ideas of health and entrepreneurship.

Fred Kent, president of the Project for Public Spaces, has a lot to say about Detroit:

Detroit is one of those extraordinary cities where what you hear and what you see are two different things. Everyone hears about Detroit being devastated by a dropping population, and it is down, but there are neighborhoods with amazing people who are doing things. There have been interventions, drawing people downtown. There are more people living downtown now and fewer people need cars to live their lives. Detroit has become one of the great destinations in North America and downtown just got a major award from the Urban Land Institute. This proves, you can do something in any kind of a city, and in a city as hard-pressed as Detroit, you can do something extraordinary.

SYMPTOMS

Detroit once had two million people, but the population has been declining for the past fifty years. At its peak in 1950, the city was the fourth largest in America. By 1990, the population dropped to about one million. In 2006, the U.S. Census reported 871,000 residents, down 8.4 percent from 2000.[1] The city is still shrinking. This outmigration is creating its own cascading set of problems. The people who are leaving are the more educated and the higher-wage earners, just the people Detroit needs to keep in order to turn around its economy.

Highly skilled people who are looking for work are finding it out of state. They put their houses up for sale. With a poor market, property values continue to decline, and the city gets less revenue from taxes as a result. The number of school-age children has declined as well, by 12,000 in 2007 alone, costing the state nearly $84 million in federal education funding. Most of those who are moving out are young, leaving an aging population, which requires a greater proportion of services. With fewer tax dollars to pay for services and a greater percentage of the population requiring them, Michigan estimates a fiscal shortfall of more than $100 million annually. As Michigan loses people to outmigration, it also loses representation at the federal level and influence in presidential elections. Kurt Metzger, a demographer formerly with the U.S. Census Bureau and now with the Detroit Area Community Information System, describes what his data depict as "a perfect storm—the education, income, young people, everything is going in the wrong direction."[2]

Detroit sits on 138 square miles (more than twice the size of San Francisco), with people spread out across the city's landscape. Despite all this potential farmland, many people cannot find decent fruits and vegetables in their own neighborhood. As the population has shrunk, properties have been left vacant and abandoned. There are thousands of acres that the city cannot serve anymore because the population has gone, and these areas are falling into disrepair. (Plate 19 is a sobering glimpse of an abandoned Detroit site.) The sidewalks are broken so people walk in the street, increasing their risk of being hit by passing cars. People are more likely to find a liquor or convenience store than a fruit and vegetable stand. In 2004, *Men's Fitness* magazine named Detroit America's "fattest city," based on the scarcity of gyms, and the number of fast-food, ice cream, and doughnut shops per capita. The magazine also found that Detroit scored poorly in availability of health care, reported levels of television viewing, and worsening commuting times.[3] Detroit had moved down the fattest city list from first to twelfth (or a bit up in terms of healthiness) in the 2009 listing.[4] Yet overall, Detroit is a city faced with tremendous challenges—declining industry, increasing health issues, and a change in its population.

Issues of Race

Michigan was an active conduit for the Underground Railroad before the Civil War. Although the arrival of former slaves did increase the city's African American population, the first real wave of African American migration was of people coming north to work in factories for such manufacturers as the Ford Motor Company. Another wave came during World War II, as factories helping to win the war needed workers. Although many African Americans did find work, they also found prejudice. It was difficult for them to find a place to live, shop, work, and play. At times the tensions and frustrations became so strong that some neighborhoods were decimated by riots, after which businesses were boarded up and houses that were damaged were left vacant.

According to Angela Reyes, executive director of the Detroit Hispanic Development Corporation, "The metropolitan Detroit area is the most segregated in the country. When people think of segregation, most would think of a city in the South, but it's Detroit and the surrounding metropolitan area. Here people have to learn how to work across racial, ethnic, and socioeconomic lines. We have to listen to and respect each other."

Lottie Spady, director of communications for the Detroit Black Community Food Security Network, sees race as a source of strength and potential:

Real Estate Developer Shawn Torrence on Detroit in the 1970s

Growing up in Detroit back in the 1970s, the city was beautiful. We didn't know that the city was segregated. The Department of Recreation would come out, providing volleyballs, baseballs, and footballs for us to play with after school. We would leave our front doors open in the middle of night because nobody had air conditioning. We would sleep downstairs and hope to catch the summer breeze.

I love the city. I didn't see a lot of crime in the Detroit where I grew up. People got along. This was a place where people sat on porches and talked to each other. If the kids in the neighborhood were doing things they shouldn't, the older folks scolded us. Then they would tell our parents, so we got in trouble again. There was a respect in the city that you don't see today.

The tenets of capitalism are based on supply and demand and someone always has to be in need in order for that to work. Meeting this need has traditionally been at the expense of some community, usually our minority black and brown population.

So, to break that trend, we need to learn new ways of becoming sustainable. We need to reach into our ancient history and remember how we as a people have found solutions and can again. We need to reprioritize. Having access to something doesn't mean it's the right thing to be doing. We need to focus our direction where it's going to make a difference. We should look at what we're eating. Habits are formed slowly and they are undone slowly. We've packaged the built environment into parcels that carry a "do not touch—do not get involved" message.

Rather than going *to* nature, or *to* the farmers' market, we need to be *in* nature, *with* nature. This is why urban farms and gardens are vital to empowering a minority population. (The Detroit garden in Plate 20 is a reclaimed urban field.) Issues of race affect the social, political, and spiritual framework of the city, along with the compounding issues of poverty and personal health. "We're unique in that we are probably the blackest community in the country," explains Dr. Calvin Trent, former director and public health officer of the Detroit Department of Health and Wellness Promotion. "We have an 83 percent African American community, and many of the issues that affect the minority communities are those issues that are resolved or assisted by public health. Public health is the safety net for many people who don't have insurance, those who live on the margins, who are plagued with substance abuse and HIV/AIDS issues. So, when we talk about communities in

need, we are talking about communities that have unique health issues. Detroit needs to be on the leading edge of public health and health care."

Issues of Poverty

Detroit outpaces the state of Michigan in all measured areas of poverty. Data from 2007 show that 33.8 percent of Detroit residents have income levels below the poverty line,[5] versus 14 percent statewide; however, this doesn't show the whole picture. When examining the real jobless rate, including part-time workers looking for full-time jobs and frustrated job seekers who have given up looking, the unemployment rate is close to 50 percent.[6]

There are lots of ideas flowing across Detroit. One way to tell whether the city is getting better is to see who is still around and who is taking financial risks and starting new businesses. It is one thing to have a series of chain stores move into an area, quite another to have small businesses owned and run by local people. Owners of long-time local businesses, such as Sheps Barber & Beauty Shop (Figure 10.1), can feel the pace of economic decline, just as they will know when the recovery takes hold.

"The effects of the economy are starting to hit here at Sheps," says Matilda Ware, Shep's daughter. "We never used to struggle. Five years ago this shop would have been filled with customers. We'd open the doors at 8 AM and ten people were already waiting. Now with people getting laid off, my weekly customers are begin to work with families when children are very young to give them hope, inspiration, and an education."

Detroit is dealing with very tough social problems, and it is going to require some tenacious measures to bring about change. This is not going to happen overnight. It is going to take some serious reeducation. In his inaugural speech, President Obama talked about the fact that there are no quick solutions to problems that are a long time in the making. Communities seeking change and revitalization will need to acclimate their residents to long-term solutions, as opposed to the quick fixes many people are used to.

Unemployment is an important health issue, not just an economic one. Unemployment is one of the most dangerous "jobs" for a man in America.[7] Link this to the environmental and social crisis, and it is clear that Detroit is in a de facto depression. Layoffs cut deep into the fabric of the community, and poor people must buy cheaper products, often with low or even negative health value. Unemployed people and families often skip preventive health care checkups and wait until a health issue is unbearable before going to the hospital. People who have given up looking for work are idle, and idle energy can exacerbate racial tensions. People don't feel safe where these tensions run high, so investors stay away from business development. In order for Detroit to move toward being more self-sustaining, new employment solutions need to be found. More than the amount of income received, doing meaningful work that contributes to the community builds a sense of self-worth, a

Figure 10.1 Sheps Barber & Beauty Shop is a shining beacon on an otherwise blighted street.

Source: Photograph from the Media Policy Center.

becoming bi-weekly and the bi-weeklies are becoming monthlies. Some people can't afford to come at all. The effects are hitting us. As the neighborhood declines and buildings are abandoned around us we can't get insurance on our building. I still have to pay my mortgage and it's getting harder."

"I believe that Detroit is one of the most impoverished cities in the nation, if not the *most* impoverished city," comments Joseph Williams, CEO of New Creations Community Outreach. "Poverty has an effect on everything—on education and incarceration. We know there aren't too many rich people in prison. In order to break poverty, we have to feeling that you do something that matters. This feeling is more important to the overall health of the individual and the community than the actual level of income received.

DIAGNOSIS

As the auto industry grew, Detroit grew along with it. As the city and America became prosperous, and to support the growing auto industry, we paved over Detroit and America. We created driving destinations, and we drove between them on wider and wider streets. Cars are made to get us from one place to another, and when they wear out, they are junked. Often we reuse the steel to build new cars, ones that are, we hope, more energy efficient and safer than our old ones.

We need to do the same thing with our communities. Eventually, communities wear out and we need to rebuild them. But this time, we must rebuild homes and neighborhoods in ways that are safe, healthy, energy efficient, and as good as they can be for the environment. There is no future in perpetuating a failed idea or in duplicating a collapse.

Detroit has lost population, wealth, and fine old buildings, and is now confronted with a built environment that no longer supports its needs. Motor City needs a new identity. This must come from individuals—from conversations around people's kitchen tables that result in fundamentally changing their homes, neighborhoods, and way of living.

Mark Kickita of Archive Design Studios understands urban design and the influence of a well thought out master plan: "The master plan guides the design direction for neighborhoods, districts, downtowns, blocks, and buildings. How we design our neighborhoods and the fabric of the city is creating culture itself. If we can provide a situation where people can walk and be active, we'll be providing a healthier community."

Sometimes saving old houses is a poor investment and is not going to work. Alter Commons is a neighborhood on the far east side of Detroit near Jefferson Avenue. It is distressed and rundown, filled with vacant lots and abandoned and dilapidated residential structures. The local Post Office keeps track of addresses that have been vacant for at least ninety days. Its numbers reflect at least a 20 percent vacancy rate in the neighborhood. People are still living in the neighborhood of course, but they need help, and this takes using what we know to build smarter homes in a price bracket that can attract new residents. Community First Development is building new, green, modular homes in Alter Commons. Mark Lewis, a partner in the venture, says the goal is "environmental consciousness, but also long-term affordability for the homeowner via reduced utility bills."[8]

The key for Detroit is in the land and how it is to be used. The land will turn Detroit around. One very positive initiative is turning open fields into places that grow nutritious and healthy food. This gives people meaningful work. The food brings not just greenery but enjoyment to people's lives.

Where is the future for a city like Detroit? One indicator might be the Eastern Market, which is both an area where food is grown and a marketplace that not only sells nutritious fresh food but also creates wealth and jobs while helping to improve the health of Detroiters.

A City in Disrepair

Detroit is an old city that has had a lot of economic turmoil. Old homes have character, but they also have issues. "One of the issues that we deal with," explains Calvin Trent, "is lead poisoning of our young children. Some of the older properties still have lead in the paint and those paint chips are available for young kids to put in their mouth." Lead poisoning affects not just the individual child but also the family and the whole community's potential growth for years to come.

If a building is unhealthy, taking it down must be considered; however, just because it is an old building it is not necessarily useless—often it is just the opposite. When thinking about sustainable building, the ideal is to use what we have, where possible. An old, abandoned, or underused building can be repurposed more economically than building a new one. We need to consider how to resurrect and reuse structures for new activities. "The greenest way of building is connecting to the past and the history of the community, bringing it back up and

creating the continuum from the early 1900s to the next century. We see this working on a number of levels and we believe strongly that using what you have can help a community move into the next phase of life," says Mark Kickita's business partner Dorian Moore.

Beyond the immediacy of the property issues, Detroit needs to think about how to get people moving. As an auto industry town, it encouraged people to drive everywhere, which created a sedentary lifestyle for many residents. The lack of physical activity has led to high rates of obesity and diabetes. It is time for Detroit to get people moving. The renovated waterfront includes walking paths. It is a start.

A City with Unused Space

As the city shrinks and the number of vacant properties grows, Detroit is rethinking how it will use available land. The city planners need a vision, and this vision is expressed in the master plan. As Shawn Torrence, formerly associated with Community First Development, explains, "it's time for *smart growth*. I think you're going to start seeing more grocery stores and restaurants, more green spaces, and parks where kids can play. Detroit can become a residential city, a city with pathways, walkways, and sidewalks. We've always been a city where we had to drive everywhere. Smarter growth can help revitalize us."

The city can bring in more entrepreneurs like Ritchie Harrison, a physical planning director for the Jefferson

East Business Association, to provide transportation and create ways for businesses to thrive. Highways that are wider than needed must go on a *commuter diet*. The city could add public transit up the middle lanes or build islands so people have jogging paths. Vacant parcels can become parks and playgrounds. Linear parks, actively tying neighborhoods together, can interconnect existing green spaces.

Interactions can also happen through new types of business. One of the promising new projects is the previously mentioned Eastern Market, which has a huge presence both as a greengrocer and as a green, sustainable farm. This initiative is using some of the city's vacant lots for urban farming, a very new push within the city of Detroit.

CURE (OR AT LEAST TREATMENT)

When we look at failed communities that are shrinking, we know these areas have lost retail stores, commercial businesses, schools, and places of worship. In addition to the loss of services and opportunities that results, there are physical spaces, holes in the community, that must be filled. *Infilling* takes vision and a path for new development. One key to infilling is determining the variety of housing. We have to decide if we want single-family houses, multifamily complexes, or mixed-use buildings with retail or commercial operations on the first floor and

residential or office space above. We also have to determine how the houses sit on their acreage, and how they interact with each other and the street. Yet another part of the work is determining where high density makes sense and where lower density is smarter. For example, creating smaller homes on smaller plots than before makes more land available for community centers like urban gardens.

The concept of the neighborhood unit was developed by William E. Drummond in 1913, and then popularized by Clarence Perry in 1929. The idea is that a neighborhood should be based on the distance that people are typically able to travel on foot to accommodate their daily needs—about a quarter of a mile. This is why major streets in a city grid system appear every half mile, so that a walk from the center of any neighborhood to a main street is only a quarter of a mile. We can use this standard to build walkable areas. When you look at Cleveland, Ohio, or Baltimore, Maryland, you can see areas that have been rebuilding and infilling. They have already gone through the process that Detroit is pursuing. Detroit has examples to learn from.

"Our goal is to shrink the wastefulness and create new opportunities from it," says Diane Van Buren Jones, a community economic consultant working to create a vision for Detroit. "The issue is what to do with vacant land in a shrinking city. There is a history of warehousing, breweries, factories, and innovation represented along the riverfront. We have assets, but what is missing is the network to get people to destinations efficiently without getting in a car.

The city should service the residents, like an ecosystem that provides all the resources needed for survival."

Home Ownership Supports Mental Health

Many of us were raised to strive toward home ownership. It is part of the American dream, a symbol of success and permanence. There is a sense of well-being when we own our home; we have a sense of security, a place to raise our family. We also want our own home so we can project our living expenses long term, without the fear of a landlord raising rents when we can least afford it.

In addition, we should want our homes to be sustainable, because when we live in these green homes we not only help the environment but we get a physical benefit. A smart green home provides a healthier internal environment. It uses paints, carpets, and other floor coverings that do not emit high levels of volatile organic compound (VOCs), and it employs high-efficiency heat exchangers with filtration. Making sure there are no toxic chemicals like lead and formaldehyde; few natural contaminants like mold, allergens, and radon; and regular fresh air exchanges results in a healthier home.

Mark Lewis describes the benefits of his company's green entrepreneurship in Alter Commons this way:

> At Community First Development, we focus urban redevelopment and neighborhood rehabilitation projects on providing green modular homes to moderate-income families and individuals. First-time home

buyers are needed to make Detroit grow again. To get people into homes they have to be economical. Modular homes allow for economy of scale. Building in a factory under controlled environmental conditions results in far less waste and better utilization of building resources. We don't have to deal with weather and other conditions that cause delay. Reducing the variables makes it increasingly possible to build efficiently so costs can be better managed.

> Typically, modular homes are looked down upon by municipalities. They think of mobile home parks and want to put us on the outskirts of town, next to the railroad tracks or nuclear waste dump, but the environment is changing.

> We see an incredible opportunity here to rebuild the neighborhood, one home, one block at a time. It's a long-term project, because there are thousands of lots available, for relatively small amounts of money. Though we are a for-profit company, we leverage partnerships with not-for-profits so first-time buyers have access to a combination of down-payment assistance and conventional mortgages. This partnership to help get people into homes is a unique facet of our business model. Even though the need is great, we recognize that it's going to be a slow process to repopulate and revitalize Detroit. Individuals have to be prepared for home ownership. That may take a year or longer, depending on credit problems, levels of financial literacy, and home buying education that may be needed.

I have always thought that the virtues of faith and hope are intertwined. For the hard-nosed business dealings that must go on during the buying and selling of homes, hope is not always a reliable strategy. However, for communities facing the bleakness of economic and population decline, hope is essential. Most Americans want eventually to own their own home. For people whose last resource is hope, the chance to have the tools to turn hope into a home can have astonishing results. With the chance to build some financial equity, a family structure can become stronger, parents can help their children stretch their wings a bit further, and the home can become equity that can be used for education or retirement.

The Eastern Market and Urban Agriculture

Eastern Market (shown in Plate 21) has been an exchange center since the 1840s. Everything from grains and flowers to fruits, vegetables, and meats has been bought and sold here and continues to be. Yet the market has also undergone a big change. Rather than simply being a conduit and distribution center between Detroit and faraway food production sites, Eastern Market is now focusing on building a local food system as well (Figure 10.2). Eastern Market staff estimate that if 25 percent of Detroit's food could be sourced locally, "it would generate nearly 5,000 jobs, create $20 million in new local taxes, and $125 million in new household

Figure 10.2 Eastern Market grows some food locally and sells it in a thriving farmers' market.
Source: Photograph from the Media Policy Center.

income."[9] Selling locally grown food reduces transportation costs and the associated greenhouse gas emissions.

Eastern Market already knows how to process, distribute, and sell food. Now it is choosing what to process by assisting in expanded food production within the city and the region. "True green," as Dan Carmody, president of the Eastern Market Corporation describes it, "is not just about energy efficiency or alternative energy development—it's about thinking more strategically and systematically about the economic vitality, social equity, and environmental sustainability of our city. A robust local food system improves the health of our most vulnerable residents and enhances our natural environment."

If we want to inspire people to grow food locally, land must be available. When communities tear down old buildings, they have a vacant lot. These spaces are like deserts. We talk a lot about the food deserts in poor urban centers. The best idea for these spaces, explains Calvin Trent, is "to grow things in this land. This land is valuable." Right now, Detroit does not have fresh fruits and vegetables in the corner stores, the liquor stores, and the gas station stores where many people actually do their shopping. So, by bringing in the idea of urban gardening, Eastern Market is changing the culture of Detroit. Urban farms can dramatically change an eyesore and a shrinking urban area, turning it from a center of disease to a path to recovery. Why shouldn't we support people in having landscapes that nurture the spirit and feed the body? Open fields and farmers' markets are sprouting up across Detroit, rebuilding neighborhoods into lively, interactive urban centers. Locally grown food is pumping vitality into the people and the economy of Detroit. This may be a model for the United States in the twenty-first century.

What I love about urban farming most is how this activity grows healthy children and cross-generational relationships. What better kind of exercise could a child do than work with parents and neighbors planting a garden?

Urban Farmer Malik Yakini, Director of the Black Food Security Network, Speaks:

My urban farm is divided into four quadrants. In quadrant two we've begun planting collard greens, one of our main cash crops. Most of Detroit's remaining population is African American and many of our ancestors are from the South. I'm a second-generation Detroiter. My grandparents moved here in the 1920s and brought with them the traditional foods from the South. These became our "home-style favorites" and the most popular cash crops we grow: tomatoes, kale, Swiss chard, and peppers. We take these crops over to Eastern Market to sell. I think what's most important about this is not the pounds of produce that we can add to the food system, but the seeds that we are planting in people's consciousness.

The model that we're creating is to show people that agriculture can be a vital part of reenvisioning and revitalizing a city. There are all kinds of links between what you eat and your overall quality of life. Diseases like obesity, high blood pressure, diabetes, and cancer shorten our lives. Increasing the amount of fresh fruits, vegetables, and whole grains [we eat] improves our overall quality of life. Good food helps us maintain a healthy weight and provides the nutrients, vitamins, minerals, and roughage that we need in order to keep our digestive system healthy. Many people say that we are what you eat, so it's important to eat right.

To make urban farming work, people need to know how to farm. Farming is more than sticking some plants in the ground. There are chemistry, biology, and geology to balance. One of the great contributions of the federal government to U.S. agriculture has been the work of the Agricultural Research Service and its extension programs that carry out the training of farmers that is part of the service's mission. Land grant universities have made major contributions to the practices of farming, yet it might be time for them to commit to urban, near-urban, and by necessity, organic agriculture—the agriculture most needed in this country.

For a long time when I worked on pesticides in California, I heard growers and the residents of nearby subdivisions argue about local issues. The newcomers would complain about the agricultural noise, diesel fumes, dust, and especially pesticide use, particularly when pesticides were being applied by crop dusters. The growers would say, "We have been growing lettuce here for a long time and have used these chemicals for years without any problem. Why should I have to change because you arrived?" The reality is that communities can almost never have *production agriculture* next door to a school or tot-lot. The other reality is that if communities put *organic farms* near housing, real estate agents and the Chamber of Commerce will describe the farms as an asset (except for the rare manure-spreading days). Not only are the members of the public increasingly demanding organic food to eat but they are also insisting on organic farmers as neighbors.

The knowledge and skills needed to make urban and near-urban organic agriculture successful need to be recovered from our senior citizens as well as propagated by current practitioners. And success needs to be more than *gentleman farming*. Communities need viable economic propositions, which is why a cadre of increasingly skilled researchers and technicians is needed. At present there is no clear commitment and no protected funding for urban agriculture, but propagation of the knowledge held by outstanding programs could be greatly improved by better funding and by cross-training, not just by academics but by grassroots programs.

This movement is a cultural shift for many folks. These little urban farms are creating joy, exercise, fresh air, friendships, and health. Engagement is changing habits and people are asking questions. They want to know how to cook what they are growing, for example. With new skills comes innovation. People are using urban farms to grow both a sense of community and healthier, stronger children.

Ashley Atkinson is director of project development and urban agriculture for the Greening of Detroit. "You have to imagine what it used to be like here," she says. "This is one of Detroit's most diverse communities, with large Arabic and Latino populations, Eastern Europeans, and Native Americans. Although very diverse, the community was very segregated." Atkinson took a thirty-acre plot of land adjacent to two schools and created a park and an urban farm on a budget of $40 a week.

We were able to raise funds for the playground equipment and teaching pavilion, three soccer fields, and the picnic shelter, basketball and tennis courts and about 200 trees. We have a 118-tree orchard. Over the past five years we've watched people come together and interact in the park and on the walking trails. We educate students through weekly programs, teaching them every aspect of gardening, from planting to harvesting. Using a passive, solar-heated greenhouse we can teach people how they can grow food year-round. Through composting we can manage our waste. We can teach adults how to grow crops and flowers to sell at Eastern Market to supplement their income. We hope they take the information back to their schools and homes, benefitting from rich and culturally diverse foods made from our crops.

Why are urban farming and Eastern Market so important? The land is putting people back to work. The crops build food security. With annual crops that supply local restaurants, people are learning how to manage small businesses rather than how to balance welfare and unemployment checks.

Fred Kent of the Project for Public Spaces says, "Eastern Market is becoming a branch of Campus Martius [a new city park], and other farmers' markets will develop in pockets around the city too, rejuvenating and reenergizing the neighborhoods of Detroit. This is part of the renaissance of Detroit. Right now, it's an attraction, but it's becoming a deeply woven part of the cultural future of the city, which is very exciting."

Campus Martius

Campus Martius began as a drill ground for militia training in 1788, named after the 180-foot stockade of the same name in Marietta, Ohio. The area was marshy and undesirable until Mayor Dennis Archer decided to make it into a public park that officially opened in November 2004 (Figure 10.3).[10] The plan was to make Campus Martius a destination for different types of outdoor activities. The park includes two stages, performances, a skating rink in the winter time, comfortable seating, markets, and food. The park is very closely connected to the buildings around it, and traffic around it has been slowed down, providing safety. "The whole place is a connected island that goes right through a part of the city," says Fred Kent.

It's sort of taken over the whole central core of Detroit. You know you're coming to Campus Martius when you're two blocks away. You feel its power, its authority. You begin to realize that you're coming into something really special. When an area that was just asphalt and nothing is made into something special in two or three years, when corporations are moving to be a part of this, it has a transformative effect. If that can happen in the most devastated of cities, and you can turn a city from something that you wouldn't

Figure 10.3 Campus Martius Park has transformed its part of the city.
Source: Photograph from the Media Policy Center.

want to be a part of into somewhere you want to be on a regular basis, it's pretty impressive.

Dequindre Cut

Our health is directly related to the environment in which we live and the lifestyle we choose. Our public health agencies know what it takes to be healthy, and the built environment can do a lot to encourage people to make the right choices. Dequindre Cut is an urban greenway, developed through a public, nonprofit, and private

partnership, offering a nonmotorized path between the riverfront, the Eastern Market, and residential neighborhoods. People can walk or ride bicycles to get fresh fruits and vegetables—exercise and nutrition in one family outing. As Mariam Noland, president of the Community Foundation for Southeastern Michigan describes it, "Dequindre Cut is a safe place for people to meet each other, walk, jog, and ride their bikes. It's a gathering place where people will connect in a brand-new way. The Cut [as it's known] is surrounded by a stable, elderly neighborhood. This area was boarded up for thirty years and now it's a center for recreation."

The Cut is an example of taking an urban feature that has outlived its use and reconceptualizing it through innovative partnerships. Once the Cut was the site of a railroad line that transported goods to and from the Eastern Market. Then it was used as a below-street-level path and became a home to graffiti artists and the homeless. In the reconceptualization of this greenway, much of the artwork has been maintained to bring the best of the past into the present of Detroit's urban culture. Few people had seen this graffiti gallery until now, and the progression of the art form is getting attention as an attraction in itself. Foundations are bringing urban artists to the Cut to further develop the art along the pathway, and reenergizing the artistic community.

As the Cut expands to its planned 5.5 miles, the opportunities for art and exercise increase. The Cut has security cameras, benches along the path, and places to rent bicycles. It promotes safety and community. People stop to look at the *graf-art* and discuss what they like. "These are things cities can make available," explains Calvin Trent, "but if we don't avail ourselves of them, it's our fault. The health department can help us change behavior and change community norms around issues of nutrition and exercise, but we have to take the hint."

The Entrepreneurial Spirit

As Angela Reyes explains:

> The essence of the American spirit is people risking everything to go someplace new, to work hard, and make a better life. This immigrant experience is very entrepreneurial. People who come here have a particular spirit and drive—a tremendous amount of courage and passion that drives them to make things better. There are people who have bought little wood-framed houses and they've turned them into big, beautiful brick homes with wrought iron fences and landscaping. They've increased the housing value in the community three or four times over the past couple of decades because people use their skills to improve life for themselves and their families.

The Argonaut Building, the old design center for General Motors, is now operated by the College for Creative Studies as a creative accelerator—a home to create businesses. The idea is to provide, at a nominal charge, the equipment people need to create prototypes for new ideas for goods and services. It's a place where you can tinker with your ideas and learn how to make them better. People help you frame your new idea in marketing terms, and funders come in search of ideas to invest in. "My daughter lives in Chicago," explains Ralph Connel, a professor at the college. "She used to ask why Detroit couldn't be more like Chicago. Now she can't believe the rebirth, the energy of the city. We are showing that we can make a difference in the quality of life."

Taja Sevelle, founder and executive director of the nonprofit organization Urban Farming, took a break from her music career to start her first urban farm.

> I was recording a CD for Sony and became acquainted with the city, how much unused land there was, and the job loss rate. I put the two together and started planting in 2005 with three gardens. Now we're planting 600 across the country. Now I get to plant, and I still sing. I love having my business headquartered in Detroit because everybody gets excited when they see the plants going in the ground. We have people coming to us crying, thanking us for food so they didn't have to choose between putting gas in the car to get to work and giving their children a meal. This is about more than supporting a food bank, which is where our food goes. It's about getting parents and children out in the garden, in the fresh air on a sunny day. What we started here is now in five countries.

Randal Charlton is executive director of TechTown at the Wayne State University Research and Business Incubator Park. He remarks that

Detroit once produced products and services that defined American culture. Now partnerships between business and universities can help entrepreneurs establish profitable companies based on health and science. We have got to use the unique advantages Detroit has to offer. We're next to another country, next to another university in that country (the University of Windsor), and there are different public health issues in these two places. Canadians have a different approach on what will work so we can do some exciting comparative studies. We're looking at safety equipment, prevalent causes of chronic disease, water issues, and technology.

Making New Habits

Detroit must get ready to pivot from *fattest city* to *healthy city*. "There is a growing chorus of voices encouraging residents to make different choices. We have to take ownership of our problem," says Dr. Trent. "Nobody is going to come to Detroit and change our situation for us. We have to take on the responsibility of addressing the issues of obesity, diabetes, and the lack of proper exercise. We have to get out there and motivate ourselves, and our institutions can help us."

"Education is very important," urges Mark Lewis of Community First Development, "educating the populace on health and helping all of us understand how important it is to get out and exercise. I think that it starts with more than the public health departments doing public service announcements about exercise [and starts also] at school and at home. If parents get their kids out and play, if they know how important it is to get out and walk the neighborhood, they'll make it happen." Michigan has physical education standards for students in grades K–12, with benchmarks for proficiency. Updated standards are accompanied by model curricula, assessments, and sample policies that schools can draw from to build a comprehensive approach. The Michigan Governor's Council on Fitness provides awards, grants, trainings, and activities to get people at all ages moving.

According to Calvin Trent:

Our whole community is fighting a good fight. I'm thrilled that so many people are picking up on these issues. It's not a matter of convincing people that we have to eat better or that exercise needs to be a bigger part of our lives. It's not a matter of convincing people that we shouldn't be shooting or otherwise harming each other. Our mayor is an athlete and we know that he's very concerned about fitness and health, and we want to bring some of those athletic concepts to our community.

So when people start eating better, exercising more and getting healthier, what will they demand of

Detroit? When we see more greenways, walkways, bicycling paths, and public transportation, what else will we see change in Detroit? Shawn Torrence says, "You'll see more people utilizing public transportation and light rail, leading to increased bus service that can get individuals from point A to point B. There'll be a lot more people walking in the neighborhoods and hopefully a *walk-to-work* environment. From that you'll see obesity decline and the dining experience won't be fast food but fresh fruit from the neighborhood grocery. As the city shrinks, I think we'll start to see healthy alternatives become commonplace and not just a luxury."

Another Opinion

Many do not share the optimism of this chapter. Nor do they share the vision of converting Detroit, whose infrastructure comfortably accommodated two million persons fifty to sixty years ago, into a low-density, garden-oriented city of a fraction of that population. They think the romantic notion of back to the garden in Detroit (where the crime rate is still ominously high) is ludicrous. Perhaps a comparison with Flint, a neighbor to the north, might be revealing. Flint is considering shrinking its overall size as a city in order to create manageable density with the remainder; it understands that basic services work well only with a certain degree of densification (albeit in a nodular form). Time will tell.

PREVENTION

We can wonder who's been asleep at the wheel as the City of Detroit has slowly declined in population and economic vitality. With each decade of decline, whose job was it to get the city to change lanes or take the next turn and develop new economies? Before we look to see whom to point a finger at, let's look to love for our families, friends, and community. When we live from love and we see something is working, we celebrate and encourage more commitment and apply resources to bolster success. And when we notice trouble brewing, we ask the tough questions. Detroit has made some headway, creating places where people can gather, exercise, and enjoy the outdoors—Dequindre Cut establishes new landmarks for healthy activity. Repurposing space for urban agriculture and local gardens increases nutrition and connectedness through community. The Eastern Market is thriving and has become a model for cities across the country. The building of economical and greener homes, and providing ways to get people into these homes, is creating hope and positive energy for those who are staying.

The economy is showing signs of improvement. TechTown is creating new businesses for the twenty-first century. Young artists and entrepreneurs are creating a viable and exciting urban core. The automobile industry is reviving. Sheps Barber & Beauty Shop is holding on with great haircuts at reasonable prices.

Detroit has been through many phases, from the industrial revolution to the Motown sound. Today Shawn Torrence says of his city:

We're ground zero. We're the New Orleans without Katrina. Detroit is caught between the industrial model and the new information age. I think Detroit is a proving ground for how cities can grow smart, and a test case for what's to come.

I had a professor who used to always talk about giving back and getting your fingers and your hands dirty by building the community, so it's something I've always wanted to do. I get up every morning refreshed, knowing this is what my job is going to be, no matter how many hours we have to put in. The fact that we are revitalizing the neighborhoods is a noble cause. What better job could you have than rebuilding and revitalizing something that some look at as not worth building?

What is even better is seeing people's reactions once we do build it. Now that is astounding.

My experience with Detroit is that it is a strong person's city, a city of courage.

Like faith, hope, and caritas, courage is the virtue without which all others are valueless.

PART THREE

BE THE CHANGE YOU WANT TO SEE IN THE WORLD

Why do you live in your community? Did you choose it actively or passively? Some people live in a community because it is where they grew up. Others choose a community because it is where they found work, where their mate lives, or where they will be close to relatives. Maybe it is where you went to school or it is near a peaceful vista. What do you like about your community? What do you not like? Have you ever taken the time to figure out what it is about your community that makes it either a pleasurable or a stressful place to live?

Earlier in the book, I introduced you to a number of places across America where people are working to make their communities healthier. You probably recognized similarities between some of these communities and yours, either before they started their work or at some point during their revitalization. Creating a healthy built environment is not about pouring energy into a place and then listening for the blast of trumpets that signifies the work is finished. This is a continuum along which we continually travel. As Elgin's former mayor Ed Schock said, "This is not something that we're going to complete in a year or two years, or even in a decade—it may never be completed, it may be something that is impossible to complete. This is an ongoing effort." However, our communities will not improve if we do not engage in the process.

Change is not always good; only the *right* change is. So, the first step is to determine what is working and what is not for the residents in your community. I encourage you to use the examples in this book to reflect upon and compare your community to others. As you examine your community through a new lens, you will see it differently, you will begin to understand that there are reasons why the community is enjoyable to live in or why aspects of it are not working. The examples presented throughout this book can help you to compare where you live with where you want to live.

Creating a healthy community takes an understanding of *community* as a concept and *health* as a way of living. It also takes the vision of a leader who can transform ideas into action. Maybe your idea is to start a new planned community, as people did in Prairie Crossing, but more likely, you already have an infrastructure to work with and the question

that lies ahead is how to use it effectively and improve it incrementally.

This final section of the book is designed as a general purpose guide to help you employ simple tools to effect real change in your local neighborhood. These chapters provide a step-by-step process for discovery and action. In carrying out the audit suggested in Chapter Eleven, think of yourself as a physician using your skills of observation to determine your patient's symptoms. In Chapter Twelve, you will identify who should be on the panel of experts to review your evidence and systematically check your diagnosis and proposed cure. In Chapter Thirteen, you will move from diagnosis into implementing a treatment for your own built environment. The Epilogue is meant to emphasize that you are not alone in your mission to make a difference. Whatever your role in your community, you can change your built environment using the resources, expertise, and tools at your disposal. Some challenges may be easy to solve—sometimes it may take just a phone call or an Internet search—and others may require a long-term effort involving the coordination of many, but the key is simply to get started. So, it is time to invite the patient into the examination room.

Chapter 11

What's Happening in Your Community?

As I discussed in Chapter Three, community happens when people connect with each other. Communities often build connectedness through gathering places like schools, places of worship, and open spaces. The built environment is not a community. It is the hardware that supports the software of community. The way we build our neighborhoods can make it easier or harder to feel the sense of community within a geographical area by encouraging interaction or hindering it. This interactive engagement between people, nature, and the built environment supports the health of a community. There is no "ultimate" healthy community, as all communities can be improved upon. Instead, all built environments lie along a continuum of degrees of supporting healthy choices. Web sites like www.countyhealthrankings.org quantify the current healthiness of individual U.S. communities.

Sometimes it is easier to identify the unhealthy factors within a community than to find the formula that creates health, but the important first step is to quantify what you have in order to make decisions on how to get what you are aiming toward.

When I was at the Centers for Disease Control and Prevention (CDC), we worked with the National Association of County and City Health Officials (NACCHO) to develop the Protocol for Assessing Community Excellence in Environmental Health (PACE-EH), which begins with defining the community and determining the community's capacity.

DETERMINING THE HEALTH OF YOUR COMMUNITY

We may know what blighted and downtrodden neighborhoods look like, but as a community moves along the continuum of health from worst case to better case scenario, it gets harder and harder to define exactly what it is that needs to change in order to bring the community to the next plateau of health. When I say that there is no "ultimate" healthy community, what I mean is that all communities do some things well and do other things that can be improved on. Remember that the modern America of obesity, inactivity, depression, and loss of community has not "happened" to us. Consciously or not, we legislated, subsidized, and planned it this way. And because we created the America we live in, we can legislate, subsidize, and plan our way to a healthier America. That is why, at any point in time, it is valuable to take an objective look at various factors across the community to determine what is working and what is not.

I wish I could tell you that every community in America is moving in the right direction, but that is just not so. Consider Black Hawk, Colorado, a town of 100 people. Following the development of several gambling casinos in the downtown area, in January 2010 the town council passed an ordinance to ban bicycle riding on the main streets within the city limits, and the town started ticketing bicyclists to enforce it. The *Denver Post* reported that Michael Copp, Black Hawk's city manager, said, "Our council looks at what they think is best for its citizens, for its businesses, which in this case are casinos, and its visitors, which are patrons that come to visit the casinos. We have had positive feedback from citizens, casinos, and our guests."[1]

This may sound like an odd situation, but the Colorado state legislature considered legislation to expand the idea—to ban bicycle riding on main streets in the sixty-four counties and in municipalities statewide. The stated reason was that such a law would be "for the preservation of health and safety and for the protection of public convenience and welfare." In fact, nothing could be further from the effect of such a law. The proposed bill was more about preserving the automobile culture.

Define Your Community

My cousins, and their parents, Uncle Bill and Aunt Sue, lived in a two-bedroom flat in Jersey City, New Jersey, by Lincoln Park. It was a neighborhood with two- and three-story family houses. I loved going to visit Jersey City. The neighborhood was always filled with other kids and stuff to do. I knew the names of many families within a block of my cousin's home and most of the kids. I have many happy memories from that time. Yet, if I were to look at that neighborhood analytically, I would have to say it was blighted. But one person's blight is another person's home. During the time when I used to visit, whatever else this community lacked, it had social capital. Sadly, social capital is like all other kinds of capital—easy to spend and very hard to accumulate.

Defining a community is not the most important part of this work, but it can be useful. In Chapter Three, I defined a community as an organizing structure. The structure might be based on square footage or the number of people living interdependently in proximity. In practice communities are often defined in terms of where they fall along a spectrum of density, land use mix, and connectivity. The continuum formed by these factors moves from natural landscape on one end to the highest density urban setting at the other (Figure 11.1).

As Ellen Dunham-Jones, director of the architecture program at the Georgia Institute of Technology's College of Architecture and coauthor of *Retrofitting Suburbia* explains, when you are looking to change the nature of a neighborhood:

Figure 11.1   The change from natural land to urban core happens across a continuum of density. Where is your community on this continuum?

Source: H. Frumkin, L. Frank, and R. Jackson. *Urban Sprawl and Public Health* (Washington, D.C.: Island Press, 2004). Used with permission of the Center for Applied Transect Studies.

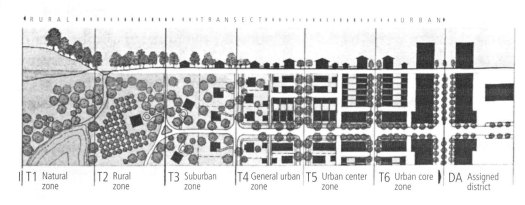

Retrofitting needs to occur at the fundamental scale of the street, the block, and the lot. You've got to retrofit urban morphology of these places to allow them to become more adaptable. The grid for New York City was laid out in 1811. They did not know what a skyscraper was. They did not know what densities the city would accommodate, but it has proven to be remarkably adaptable. I think part of our concern is that we look at most of suburbia and we don't think it is as adaptable as it is. What we're seeing in the research is that it takes a pretty significant change to instill connectivity, to reinstall the green space that allows for a much healthier living pattern.

The way our neighborhood looks did not *happen to us*; we thought it into existence. When we built a road, we had cars in mind. We were not thinking about children or the elderly. We humans adapt quickly; we quickly come to think that things have to be the way they are currently, and we can persist in that until we are introduced to change. It does not *have* to be the way it is.

Currently dean of the University of Washington School of Public Health and former director of Environmental Public Health for the CDC, Howard Frumkin has long been concerned with what it takes to create and maintain healthy communities:

With sprawling areas where people are highly dependent on their cars, we see more obesity and more of the diseases that flow from obesity and sedentary lifestyles. We see higher rates of fatalities from car crashes. It is a simple epidemiologic fact. The more time we spend in a dangerous microenvironment, and the car counts as one, the higher the risk of something bad happening. Every time we drive a distance of 70 miles, we run a one in one million risk of death. The air pollution our vehicles emit kills more Americans than the crashes. We see a direct correlation between higher levels of air pollution and ozone levels where there are high levels of automobile traffic. Those unhealthy features of the community flow directly from design decisions.

The "What do we do?" question is a great one. We need local evidence—not just national—because we need to be able to tell home buyers, planners, and policymakers about the health benefits of locally appropriate, good design. We need policies that encourage good design and that means policies that point us toward a better mix of transportation options and different land-use patterns than have prevailed for the last few decades. We need an educated public because people need to know the impact of their choices about where to live, where to work, and how to travel. The public needs to know the benefits that can flow from good design, and we need the market to meet our needs for healthy places.

People are choosing more and more to move back into more traditional neighborhoods, ones that are not so dependent on cars. Often it is a matter of convenience and economics, but it is good for health. A series of factors are converging to help us move toward a healthy community design. The nice thing about healthy community design is that it is not just something we do for health reasons. We also do it for environmental, economic, lifestyle, and livability reasons. When a multitude of benefits converge on a single set of decisions, the results can be life changing.

Advocate for Legible Buildings

Way-finding is a health issue. The medical center where I work has a subpar working environment. Even though it is called the Center for Health Sciences, it is designed to be the opposite. Each floor appears identical to all the others, with low ceilings, fluorescent lights, unreadable floor maps, and a numbering system so incomprehensible that it puzzles even my friend and biostatistics genius Professor Ron Brookmeyer. Stairwells are randomly placed, poorly marked, dirty, and without windows, even when placed against outside walls. Maddeningly, the stair doors are locked on some floors. Stair doors also have no windows, so building occupants regularly smack each other when entering the stairs. The elevators, in contrast, are easy to find—and often crowded with health professionals whom I think should be walking. For me, it all recalls the classic bad dream scenario where you are walking down a dark hall and all the doors are locked.

My office door has a window in it. The old door fell apart, so I asked for a window in the door when it was replaced. I thought, if there is a power failure, and the hallway is pitch-dark because sunlight from the office windows cannot shine through the office doors to light the hallways, my colleagues and I will not be able to find our way out of the building. I want to make this space better for people. That window in my door is actually there for health and social reasons.

Most living things must be able to figure out how to get to and away from places. When we lose our way-finding ability, we are at the least very uncomfortable. To build structures that do not enable and enhance our ability to go where we want to go does not make sense. When we walk into the ruins of ancient cities in countries such as Greece, we find that we know exactly where we are going, even if it is our first time there. A lot of thought was put into those buildings and the city plan. Directions are confusing only when the designer or architect wants you to be confused, for example, when defensibility from attack is important.

Consider Light, Balance, and Function

Recently Harry Wiland, a film crew, and I went to Syracuse, New York, and talked with Edward "Ed" Bogucz, executive director of the Syracuse Center of Excellence in Environmental and Energy Systems. Ed Bogucz's organization partnered with the City of Syracuse, Syracuse University, and a coalition of local businesses to offer a competition for architects to design healthy homes for vacant lots in urban neighborhoods needing to be infilled. One of the three winners, designed by Cook + Fox Architects, applies a concept called *biophilia*, an inherent need in humans for interaction with nature, through its use of lighting. "There's a screen on the outside of the house designed to mimic the dappled light coming through a canopy of trees. There is a series of light tubes on the inside that bring in a light that mimics light in natural settings. Broad

windows provide opportunities for occupants to interact with the outdoors," explains Bogucz. The use of light does not have to be expensive or complicated to be comforting.

As I discussed in Chapter One, buildings used to be about solidity, usefulness, and beauty (*firmitas*, *utilitas*, and *venustas*), the three quintessential elements of architecture identified by Vitruvius around 50 BCE. These are important attributes of a healthy human body as well, as Leonardo da Vinci illustrated in his drawing *Vitruvian Man* (Figure 1.3 in Chapter One).

Libraries built during the Carnegie era, about 1880 to 1930, were iconic and known for their beauty. Great efforts were made to adapt them to the local architectural vernacular. So too, schools built before 1935 were beautiful buildings that reflected the community's sense of the importance of the work happening within them. As I've said before, the built environment is social policy in concrete, reflecting what we think is important and what or who is not. Buildings can make people feel more or less valuable. In the 1940s and later, even buildings intended for very different purposes began to look alike and communities accepted this utilitarian approach. Until very recently, even school buildings and prison buildings were often indistinguishable from the outside. Buildings tell people what society thinks of them. What do the buildings in your community tell you about what was important to those who built them?

Daniel Solomon, an architect in San Francisco, is considered one of the fathers of New Urbanism. In a recent presentation, he astonished me by explaining his theory that modernism was a response to germ theory. We used to build rococo architecture—ornate, highly detailed, and decorative. Solomon suggests that we applied the antiseptic and sterile methods resulting from germ theory to architectural design; we went from overdecoration to sterility. Germ theory has conferred great benefits on us, and we do need our places and our bodies to be clean, but even our own bodies need bacteria to function. I remember taking care of an infant with a severe infection whom we were treating with antibiotics. One day he began to bleed profusely and started to look like a blueberry muffin. We quickly realized that the antibiotics had killed the healthy bacteria in his gastrointestinal tract that supply vitamin K, an essential clotting element. We immediately gave him doses of vitamin K to restore the right levels. Once again the lesson was clear—*health comes in the balance*.

If we all need various kinds of balance in our lives, we also need a sense of place. When we realize that there must be 50,000 four-way intersections in America that consist of a gas station, a fast-food franchise, a car-related business, and a bank, one on each of the four corners, we also realize that most of America has been designed by cookie-cutter and is indistinguishable from place to place, the antithesis of the Carnegie libraries. If we were dropped into one of these intersections we wouldn't know where we were, as there is no sense of *place*. The former mayor of Milwaukee, John Norquist, now head of the Congress for New Urbanism, once said

to me, "We have spent the last sixty years building things and places that no one bothers to take a picture of." No one wants this for his or her community.

Robert Davis, who built Seaside, a New Urbanist community in Florida, walked me through his town, pointing out various structures. Of one he said, "That's a temporary building, but of course, all buildings are temporary." It struck me—*all buildings are temporary*. To the degree that we can take ownership of our landscape and our places, we can make changes in our community so that the neighborhood reflects our values, vision, and personal style. This is about change and what is possible.

As we heard in Chapter Six, even though Charleston's mayor Joe Riley emphasized the importance of maintaining historic downtowns, he also described a vision for creating a publicly accessible waterfront, with walkways, parks, and picnic and recreational areas for all to enjoy. Part of Mayor Riley's legacy is visionary design. He also helped to start the Mayors' Institute on City Design, an opportunity for mayors to sit with top-quality urban designers and benefit from that input as they decide how to redesign their communities. Mayors put maps of what they want to change in their community in front of the group, and everyone provides feedback. At a recent institute session I attended, Mayor Ann Johnston from Stockton, California, asked the group what she should do about open and abandoned fields in a highly visible and semi-industrial area near Stockton's port. Sam Adams, the mayor of Portland, Oregon, said, "You know what I'd do? I'd get a farmer to plant there. Everyone would be happy if there was a corn or pumpkin field near town. No one would know it is 'lost' land." This is simple, inexpensive, and useful. What a great idea!

Seek Connectivity

A well-designed, healthy, wholesome, sustainable neighborhood has connectivity, so that it is easy to get from one place to the other. It has mixed use, meaning that residential, commercial, educational, recreational, and other uses are relatively close to each other. It has good pedestrian and bicycle infrastructure, meaning that people have a choice in ways of traveling other than by automobile. It has pretty good density. We are not talking about New York City for everybody, but communities need enough density to support transit systems so that commercial and educational facilities can be sited closer to where people live. (Plate 25 illustrates street life in an older Los Angeles community.) Sometimes we forget that at least one-third of the U.S. population doesn't drive.

Now, all of these features, and some others, functioning together, lead to certain changes in the way people travel and live their lives. Travel demand is decreased in a well-designed neighborhood. If we need a quart of milk or a loaf of bread, we should not have to drive three miles to get it. We should be able to walk a block or two. Notice all that results from this: we are driving less, burning fewer fossil fuels, and replacing driving with some

routine physical activity. We are creating less air pollution and are not at risk of being in a car crash when we're not in the car. A range of health benefits flow directly from this idea of having a community designed for walkability.

Poor design, in many ways, does just the opposite, and that characterizes a lot of what we have done in this country over the decades since World War II. We have built at a low density, so the places where we live, work, and play are spread over big geographical distances, resulting in a heavy dependence on automobile travel. We have built for a low land-use mix, which means that residences are in one place, commercial properties are in another, and employment is elsewhere—with long distances between all of them. (Plate 22 is an example of suburban isolation.) Yet more and more, as the scientific evidence accumulates, we have very good indications, based on science, that people travel differently, live differently, and live in more health-promoting ways when community design follows healthy guidelines.

Different people have different preferences, and no one person or viewpoint should be dictating what kinds of neighborhoods people live in. We really need to aim toward providing choices, so that people can find the neighborhood that feels best to them. For example, I live in a neighborhood that I love. It is a neighborhood with trees and with good sidewalks. With a five- or ten-minute walk from my house, I can get to stores, so when I need to do my basic shopping I can do that, and because traffic volume is relatively low, I can also bicycle safely when I want to. I use the bus system to get around

Los Angeles, and there is a bus stop just a few minutes from my house. In fact, I wrote much of this book while going back and forth to work with my collaborators at the Media Policy Center in Santa Monica. Many people have forgotten—or have never even had the chance to learn—how much work you can get done when you are not behind the wheel.

At the same time, our frequent feeling that different groups of people always have different goals and vastly different concerns may often just be wrong. Recently, I was asked to give two presentations—one in Oakland, California, to a group I would describe as 60 percent low income and people of color, and one in Palm Springs, California, to a group I would describe as financially comfortable, older white folks who spend as much time in a golf cart as in a car. At the end of my talk, I got the same comments from both groups. People said, in effect, "I don't like the way we have constructed the world." Grandmas in both places said the same things: "Kids don't have fun"; "I was able to come and go to the playground and to visit my friends when I was a child. Now everything requires a car ride"; "Food used to be interesting." In my personal relationships I have noticed the same thing. My most conservative and my most liberal friends often say the same things: "I don't like the values we are giving our kids"; "I'm very worried about the trajectory of our country."

The shelf life of adults is short. We have to focus on young people who can continue the work of change. It is going to take a generation or more to turn things around. We benefited from some things that our grandparents and great grandparents put in place, and we owe this same opportunity to our children.

I recently met Richmond, California, resident Toody Maher, who is taking this thought into action. When Toody Maher was young, she was a successful entrepreneur. Now that she is a bit older she has looked around her city and has come to the realization that it needs more parks. After purchasing a set of land parcels, she has begun to build a park. I am inspired by her heroism in trying to turn her neighborhood around. Richmond may be deeply blighted, but with people who have vision, as Maher does, communities will turn around. She picked the needs of a child as her focus. I like the way this speaks to the idea of looking forward.

CONDUCTING AN AUDIT OF YOUR BUILT ENVIRONMENT

An audit is an objective survey of an area, designed to collect data that can be analyzed. There are many types of audits. For example, an energy audit will tell you, by categories, how much energy is used within a structure. By carefully collecting and then examining these data, we might uncover where energy use is above expected levels, and we might find specific areas of waste related to the building, the equipment, or the usage pattern. If these wastes are corrected, quantifiable electricity savings can be realized.

A community can be audited for its health as well. You might follow the model used in Part Two of this book, thinking of yourself as a physician whose job is to hear about symptoms and collect data so a diagnosis can be made. At that point, specialists may need to be called in to identify a cure and a program of prevention, but the place to start is to get an initial snapshot of the patient. When a patient comes into the examination room, you

Figure 11.2   Huge roads don't work well for the number of cars we have and don't work at all for nondrivers, so what are we doing building more of them?

Source: Graphic by Scott Izen for the Media Policy Center.

can see the person, get a sense of his or her energy level, you can do some basic tests, and you can listen.

I would like to challenge you to conduct an audit on the health of your built environment. Start by simply walking around your community and taking a fresh look at the world around you. *What* you look at in the community and *how* you look at it is important. Think of using a variety of lenses that allow you to focus on different aspects of your neighborhood. For example, consider the elderly woman I described in Chapter Two, whom I saw walking along a seven-lane highway in Atlanta, Georgia, in the heat of the summer (Figure 11.2). The problem was not that she lacked a car or that she had gone shopping on a hot day but that there was no safe way for her to get from the store to her home with her packages. She, and all the other nondrivers in the community, had been left out of the planning process. The systemic development of the built environment did not include meeting her needs.

Decide what it is that you want to examine closely (see the list of possible areas of focus at the end of this chapter). Look at the evidence you have gathered, and ask yourself why things are the way they are.

Marice Ashe, executive director of Public Health Law & Policy, in Oakland, California, offers this insight: "The modern way of looking at community organizing to improve health is not to look only at gaps but to also identify assets. Don't just look for what's missing, but look for what's there. What's working? What strengths can you bring to bear on the areas where you want to make improvements? Asset mapping is usually the first step in engagement with the community."

So as you travel through your neighborhood, also notice the strengths of your community. This includes the community aid organizations, schools, faith-based groups, charities, medical facilities, economic resources, historic landmarks, natural sources of beauty, areas for physical activity, and places for people to meet. You might identify the location of these assets as an overlay on a map of your community so you can see the proximity of different assets to one another and to other elements and can consider how they might interact or overlap in services.

Now think about what more is needed to make the community healthy. Identify the institutional barriers and the leverage points that can be applied to change those institutional, social determinants of health. There are plenty of examples throughout Part Two of this book that you can use to get started.

Every community has a distinct set of assets. What are the best ways to capitalize on them? Keep asking yourself what might be changed to make your community a healthier place to live. You probably already know most of what you are going to find. If your community is soulless, it will not come as a surprise to you, but you might be in for a surprise if you put a clock on the time it takes you to reach a grocery store or a public park.

Write down the observations you make and the thoughts you have about what you are seeing. Take pictures with a digital camera; the images can convey a lot of information when you are creating an audit report and sharing your ideas later. Talk to your friends and neighbors and find out what they think. If there are active community organizations in your area, contact them and share what you have noted, and find out what they are working on. Remember, all smart interventions start with a good asset and deficit assessment.

POSSIBLE AREAS OF FOCUS FOR A COMMUNITY AUDIT WALK

Housing. What is the variety in the housing stock in your neighborhood? Map the single-family homes, apartments, condos, farms, and so on. About how many bedrooms are in each dwelling? You do not have to do a survey—just estimate based on what you know and can see. The greater the variety in housing stock, the more likely it is that the community will possess a mixed socioeconomic and age base. How is housing funded? Are there programs to provide employer-assisted housing and to ensure affordable housing to low-income residents, students, the elderly, those with disabilities, or others? What properties are for sale and are they in foreclosure? You might take a look at real estate listings for your neighborhood to get a picture of the homes available.

Brownfields. Map the locations of abandoned, idled, or underutilized commercial or industrial sites. Are there empty commercial buildings with for lease or for sale signs on them? These could indicate the beginnings of blight. Oftentimes this information is available online.

Open spaces. Where is the open space, and how is it used? Are there neglected vacant lots or blacktop parking lots? Identify the locations of parks, walking paths, amenities for physical activity, and natural preserves. Are the areas well lit? Do they have trash receptacles, benches, and drinking water? Is there security? Computer-generated maps can be helpful here as well.

Commercial ventures. What are the sources of income in the community? Are there cooperative models, franchises, or family-owned businesses, and where are they located? Who works at these establishments? Are the employees local to the community or do they commute from long distances? Make a list of the businesses in your community—those that are within ten minutes by car or thirty minutes by foot—www.walkscore .com may be helpful here. How many people can be employed at living-wage jobs by these businesses? Are these businesses environmentally friendly?

Public transportation. Where are the bus, ferry, subway, or rail stops? How far does someone have to travel by foot to get to needed public transportation? Can people easily transfer from one form of transportation to another? Can they travel to and from needed services easily and safely? Is the cost of this transit reasonable given the economic level of the residents?

Roads and highways. Where do people park their cars, and what is the condition of the roads they travel on? Note the proximity of heavily trafficked roads to residences and schools. Determine whether cars or trucks idle alongside residential areas.

Education. Identify formal and informal educational settings and the number of students taking advantage of their offerings. For formal schools, note rates of absenteeism and truancy, graduation rates, types of specialized programs offered, and whether their employees live in the neighborhood or commute to work.

Cleanliness. How is your community maintained? Is there trash in and along the streets? Are the storm drains clear of debris? Are trash cans overflowing or do they smell? Is there animal waste, or do people clean up after their pets?

(Continued)

Access to food. Where do people go to get their food? Where are the corner markets, supermarkets, big-box stores, farmers' markets, and community gardens? What kinds of food can be purchased or traded at each outlet? Can local residents find nutritious, affordable food? What kinds of restaurants are in the area? Make a list of retail food outlets within ten minutes by car or thirty minutes by foot. How many of these food outlets offer fresh fruits and vegetables, quality meats, and staple food products for purchase? How many fast-food franchises and how many sit-down restaurants are there? How easy is it to get alcohol for immediate or later consumption?

Access to medical care. What medical facilities and clinics are in the neighborhood? What services are provided, and what is their success record? How are patients treated, and do providers speak the language of the local residents? How quick is the emergency response time, and does this speed come in part from high-speed roads with high-speed turns? Often these accommodations make a place less safe, not more. Our streets are often unnecessarily wide to accommodate occasional emergency vehicles, even when homes are only one or two stories high and hook and ladder fire trucks are unlikely to be needed. Are there programs to meet the needs of special populations? Are preventive as well as emergency services available?

Historic character. Identify buildings and organizations that reflect the historic character of the community and note their current state. Restoration of buildings can stabilize the economic base of an area.

Walkability. Use a resource like www.walkscore.com to see how far it really is from your street to essential services. Then try walking to those services—what do you notice? Are there sidewalks, crosswalks, or obstructions along the walking path? Do people park across crosswalks, and do the police ticket them for doing so? Do homeowners trim vegetation back from sidewalks, and does the city enforce that requirement?

Physical activity. What amenities are available to people living in the community? Are there basketball courts, baseball diamonds, bowling greens, golf courses, walking tracks, swimming pools, exercise circuits, fitness centers, and so forth? Are they privately owned or publically operated? Are they clean and safe? Are outdoor facilities lit in the evenings? Are school recreation areas available to the community before and after school hours?

Air quality. How is the air quality where you live? Do you have smog, thermal inversion layers, dust, or fumes from industry? Are industrial vehicles coming from elsewhere and going elsewhere passing through your neighborhood? What strategies and resources exist in the community for improving air quality in homes, schools, businesses, and outdoors? Identify those structures that create air pollutants, and learn whether the pollutants are mitigated at the source or elsewhere. Talk with health professionals about rates of asthma among the people in your community.

Water quality. What is the quality of the local water used for community irrigation? What is the quality of the water used in homes and workplaces? Are there stagnant pools around? Do any ponds smell like rotten eggs? What strategies and resources exist to ensure water quality and conservation?

Land and soil. What strategies and resources exist in the community for improving the quality and health of the land and soil in residential, commercial, and industrial areas? Identify possible sources of soil and land impurities, such as lead, and learn what is being done to mitigate impurity levels.

CHECKLIST OF POTENTIAL ISSUES

Use this checklist, or one that you have made, as a guide when you are looking for diagnosable symptoms in your community.

Housing

❏ Are there homes for individuals, couples, and families?

❏ Are there homes that meet the needs of diverse populations (the elderly, persons with physical or mental disability, students)?

❏ Is there subsidized or employer-assisted housing?

❏ Is there a mix of rented homes and owner-occupied homes?

❏ What are the characteristics of properties for sale? Are they all in a particular area? Are there a lot of foreclosures?

❏ Do the homes in your area smell musty on the inside?

❏ Are there homes in substantial disrepair?

❏ Is there incidence of asbestos or lead-based paints in homes in your area?

❏ Are buildings in your community generally well insulated?

❏ Is your community integrated across all walks of life, or is it segregated by income, ethnicity, or age?

Brownfields and Open Spaces

❏ Are there abandoned, idled, or underutilized commercial or industrial sites?

❏ Are there a number of commercial buildings for lease or sale?

❏ Does your community provide access to parks and recreation facilities? Is there adequate open space for play, exercise, and the enjoyment of nature? Is there a baseball or soccer field that children and adults can use?

(Continued)

❑ Does your community support a dog park where people can exercise with their pets and interact with other pet owners? Can the elderly and children join in and interact with the animals even if they do not own a pet themselves?

❑ Are there amenities in the open spaces, including water, trash receptacles, and benches?

Commercial Ventures

❑ Are the essential services you need to run your home available within walking distance?

❑ Do you feel welcome at local businesses?

❑ Is there a defined community center where people can interact?

❑ Do the public buildings in your community have enough fresh air and sunshine to be attractive?

❑ Have businesses, public buildings, and personal homes been prepared for earthquakes, floods, or other perils that exist in your area?

❑ Are businesses owned locally, or are they franchised?

Public Transportation

❑ Are bus, ferry, subway, and rail stops convenient?

❑ Can people transfer between public transportation methods to get where they need to be?

❑ Are public transit options safe and affordable?

Roads and Highways

❑ Are roads and highways in good repair?

❑ Are there buffers between heavily trafficked roads and pedestrians, schools, and residences?

❑ Do trucks idle alongside the roadways?

❑ Does your community have complete streets (ones that accommodate all users)?

❑ Are there sidewalks, and are they in good repair?

❑ Does your neighborhood have adequate and safe provisions for bicycling and public transit?

❑ Are there several routes you can take to essential services (so that walking does not become boring)?

❑ Is traffic congestion or high-speed traffic an issue in your community?

❑ Are the streets and public spaces safe for children and the elderly?

❑ Are the streets and sidewalks clean and free of debris and obstruction?

❑ Is your community a heat island? (Do vast expanses of paved land mixed with buildings that reflect heat back onto the street raise the temperature in living and working areas, making it much higher than the temperature in the surrounding landscape?)

Education

❑ Is there a public library, an educational facility for adults, or a community center?

❑ Do schools offer a variety of programs or activities for students after traditional school hours and on weekends?

❑ Do school employees live in the neighborhood?

❑ Are your schools good places for children to learn and grow? Do they provide adequate access to fresh air and sunlight?

Cleanliness

❑ Is there a widespread problem with vermin (such as rats, mice, or cockroaches)?

❑ Is trash placed in receptacles? Are they overflowing? Do they smell?

Access to Food

❑ Do you have ready access to a supermarket with fresh foods that promote health or to a farmers' market?

❑ Do local restaurants serve real food, or deep-fried fast food made up only of low-grade meat, fat, carbo-hydrates, and sugar?

(Continued)

❏ Can you travel by foot to food outlets?

❏ Is there easy access to alcohol for immediate or later consumption?

Access to Medical Care

❏ Does your community seem to you to have a higher than average incidence of depression and mental illness?

❏ Are there medical facilities and clinics in the community?

❏ Are preventive and emergency services available?

Physical Activity

❏ Are there public facilities where people can get physical activity?

❏ Are there commercial facilities (such as fitness centers) within walking distance?

❏ Are there a variety of amenities geared to different age groups and fitness levels?

❏ Are the available facilities and amenities clean and safe?

Air, Soil, and Water Quality

❏ Are there centers of pollution (such as blighted parcels of land, dirty rivers, or active or abandoned buildings)?

❏ Are there many mobile sources of air pollution (such as large trucks), and are there many point sources of air pollution?

❏ Are surface water sources and wetlands in your neighborhood polluted? Do they support wildlife?

❏ How does water enter and leave your neighborhood? Is there runoff, and what could be in it?

❏ Does vacant land in your area grow weeds or anything green, or does it not support plant life?

❏ Is there a chronic disease or ailment affecting people in the community?

❏ Are air, water, and soil monitored regularly?

Chapter 12

Who Are the Players?

Michael Josephson of the Josephson Institute tells this story:

A young boy was walking with his father along a country road. When they came across a very large tree branch, the boy asked, "Do you think I could move that?"

His father answered, "If you use all your strength, I'm sure you can."

The boy tried mightily to lift, pull, and push the branch, but he couldn't budge it. Discouraged, he said, "Dad, you were wrong. I can't do it."

His dad said, "Try again."

This time, as the boy struggled with the task, his father joined him. Together they pushed the branch aside.

"Son," the father said, "the first time you didn't use all your strength. You didn't ask me to help."[1]

There are many things we cannot do alone, but that does not mean we cannot get them done. We are all surrounded by resources that we can apply to help us achieve our goals. Sometimes we fail to ask for help; we think it is a sign of weakness to admit we need a hand, or we do not even think about asking others. Whatever the reason, it is important that we learn to use all our strength. Just as we should be willing to help others, we should be willing to ask for the help of others. This is an important part of the class discussion when my students and I talk about the ethics and strategy of implementing good design (Figure 12.1).

As we address the major issues in front of us, we have to keep what is important in focus. We do not have to think about making life a lot different. But we do need to work to bend the arc of change toward health.

Making change happen in your community involves communicating ideas and creating partnerships. The *players* in this process are the people and organizations with a vested interest in either making change or keeping

Figure 12.1 Richard Jackson leading a class discussion at UCLA.

Source: Photograph from the Media Policy Center.

Remember

Change is a constant.

Change takes time.

Timing matters.

To make the *right* change, involve others with expertise and resources in the process.

Talk to everyone.

Build alliances.

Stay open to new directions and working with new people.

Do the research.

things the way they are. When it comes to changing the built environment, the major players include individuals, companies, and agencies that live, work, provide services, and play in the given vicinity.

Change—especially good change—usually takes time and a fair amount of patience. For example, as Kevin Krizek, professor of planning and design at the University of Colorado and director of the Active Communities / Transportation Research Group, points out: "Boulder did not retrofit the city overnight. There are a number of things that they've been chipping away at over time."

I am going to let you in on an important strategic insight. If you want changes, and the eleven people you are working with also want changes, showing up at an elected official's door with a dozen demands from twelve individuals guarantees frustration and failure. And the elected official is relieved; she doesn't have to do anything. If all twelve of you agree on Priority 1, and advocate only that priority, and do it repeatedly, you are far more likely to succeed. Then you can move to Priority 2. Fundamental change to the culture of a community usually happens in stages, and it must be strategic, with hard-nosed priority setting. And the setting of priorities must include many interested parties, including (sorry for the bad news) the naysayers and grumps. You may ignore them later, but they often bring important advice to the decision process—and if you can't get your allies to buy in, you certainly won't get your adversaries to do so. If you manage the strategy process well, leading and working to achieve the goals becomes easier. When you have a record of success, the next challenges become less formidable. I have always liked this saying: "Power is something other people think you have."

Tracey Winfrey of the Boulder Public Works Department reflects that "Boulder has been lucky in that we have had political support for our initiatives on an ongoing basis. I think some communities are just emerging to a point where they have advocacy groups, boards, and city councils saying, 'Hey we could be doing this, let's do it,' and then it becomes institutionalized within the government and its approach to doing business. Only then do you start to see progress. The challenge is that building momentum takes time."

So, you've finished walking around your community (see Chapter Eleven), and you've recorded your observations. Now it is time to brainstorm a list of questions, and it is also time to begin thinking about the most difficult one—how will you know when you've succeeded in identifying one or more issues where change will make a real difference in people's health?

Heather Kuiper, one of my graduate students at Berkeley, makes this suggestion:

> The first thing is to start where you are, look at your own life and lifestyle, and how it could be different. If you're driving to the store, why are you driving? How can you get to the store without driving? If you cannot let your kids walk to school, why not? The first step to answering the question is to understand it. That means to look at it, examine, and measure it.

As people become serious and gain a deeper understanding of the issues confronting them, a whole fun and exciting world becomes available to them.

Maybe you think you know what the major issue you want to address involves, but keep listening. You could be mistaken. Wendel Brunner, director of public health for Contra Costa Health Services, offers this example of working with an evolving issue:

> In public health, we are compelled to act in an environment of uncertainty. When I first came to Contra Costa County, the rate of lung cancer was 40 percent higher in the industrial area than the nonindustrial area (as measured by the white male population). We first thought that this very high incidence of lung cancer was due to the industrial pollution in the area. The most logical "fix" would be to close the industries in the area to reduce the pollutant levels, so we started looking into actual airborne toxins. What we found was that the increased levels of lung cancer were due to smoking, not industrial air pollution. We had to start asking a different question—why do [white males] in low-income and minority areas smoke disproportionally [more than white males in] the surrounding areas. We had to look at the environmental factors that promote smoking in this population. What we found was both a relationship between the built environment and human behaviors, and the lingering legacy of cultural identity.

FINDING YOUR STAKEHOLDERS

Many communities have established teams to address community issues, and it is usually easier to engage an existing network than it is to start a brand-new one. Chances are that you can connect with an existing team whose members know how to get results. You just have to find the right one.

Nsedu Witherspoon, executive director at Children's Environmental Health Network, in Washington, D.C., is a master at helping people connect to solve community issues. She remarks that in her area of endeavor, "What's exciting is there are county, state, and local government resources, nonprofits, for-profits, you name it, who, even if it's not their number one mission, a piece of their mission is thinking about children's vulnerabilities. There are a lot more resources available today than even five years ago, and we know who to turn people to. A lot of people call us and we end up being a mediator, giving them all kinds of localized, regional resources, and groups to follow up with."

Wendel Brunner echoes Witherspoon, noting that "county public health offices have a variety of resources. While each organization is different, there are a lot of resources available electronically, and staff will help point you to the appropriate ones."

So with all that in mind, think of your team as an advisory body—a group of people who can shed light and a variety of new perspectives on the issue you're wrestling with. You may be the one making the final decision on direction, but it is an informed decision. Let's take a look at some (although surely not all) of the stakeholders you will want to consider for your team.

Governments

Is the problem you have identified subject to government regulation and control? Government is, or should be, about solving problems and protecting public health, but sometimes agencies and administrators need a shove in the right direction. As Ed Schock, the former mayor of Elgin, Illinois, commented, "I think the citizens, not just in this country, have been on the leading edge of the sustainability movement, and government at all levels has been slower to catch up, but I think we are catching up."

Government officials, both appointed and elected, set policy and make critical decisions, often affecting the built environment and public health. Howard Frumkin, former head of CDC's National Center for Environmental Health and currently dean of the University of Washington School of Public Health, points out that "at the policy level, decision makers have a very important role to play. From the local zoning board to the federal Department of Transportation, policies can incorporate healthy and environmentally friendly elements, and it's very important that policymakers are alert to make those decisions."

When you identify a problem in your community that will require government cooperation to fix, make sure you find the right level of government to effect the change. In other words, the fact that something is happening inside the city limits does not necessarily give the city government the authority to change it.

For example, a friend of mine attended a community meeting about noise problems in a particular neighborhood. This neighborhood had, among other things, an interstate freeway, rail lines, and arterial roads heavy with commercial truck traffic, and the community was also in the flight path of both a large international airport and a National Guard air base. The meeting was called and chaired by the city council, but there was little the city could do about the issues raised. For each of the noise sources, a different government agency held regulatory authority. For instance, each railroad train passing through the neighborhood is required to blow its horn seven times as it approaches road crossings and other features, no matter what the hour of the day or night. That regulation was created and can be altered only by the Federal Railroad Administration, which governs rail traffic in America. The city has no say in that process; nevertheless, the city government is a powerful ally and can get the attention of a federal agency more easily than an individual, concerned citizen can. Similarly, the airport hours and flight paths are under the control of the Federal Aviation Administration and the Department of Homeland Security. The Air National Guard is governed by the Department of Defense, and the interstate

highways are under both state and federal control. Even the truck traffic on city streets is governed by Department of Transportation standards. The city was able to promise additional speed limit enforcement—but that was all. Sound mitigation was minimal.

So, what level of government do you turn to? Where should you start? Before you take your concerns to city hall, do your homework, and make sure that the level of government you're contacting has the authority to help you. Almost all government agencies have posted their regulations on their Web sites. Online is a good place to start.

The Municipal Government

Municipal, or city, governments are critical stakeholders for decisions on land use and planning. City planners often need to be reminded of the impact that current land-use policies relating to the built environment have on public health. These are the people who determine whether there will be sidewalks, whether the local parks will be maintained, and whether there will be a fast-food outlet or a farmers' market on the corner lot. Land-use planning is, at its core, about zoning, and the outcomes of these choices affect the locations of transit stops, the availability of transportation facilities, whether or not food deserts (where there is no local grocery store) exist, and whether or not a community has high-density housing and retail development.

Dave Kaptain, the new mayor of Elgin, Illinois, and a former Elgin City council member, recalls that "I served for six years on the Elgin Planning Commission and as part of that we reviewed annexations and new businesses that came to Elgin. I started to appreciate the value of economic development and how that revenue, the businesses, and the people that are associated with them help form the basis for a community."

Ellen Dunham-Jones, coauthor of *Retrofitting Suburbia,* offers this observation about city governments:

Figure 12.2 Counties oversee permits for zoning and road construction.
Source: Photograph by Joseph Sohm. Used with permission.

Certainly one of the things a city can do to be proactive and try to incentivize redevelopment is to re-zone. Many zoning laws still in effect are outmoded and, in fact, detrimental to the public health and well-being of the community. During the boom years, a lot of planning departments were reactive, responding to developer requests. Now they're in a proactive mode, asking, "What is your vision of the future? What do we [as a community] want to become?" Oftentimes, it comes down to which is a higher priority: quality of life or deference to the automobile.

The County Government

Counties are a critical layer of government. Their functions are mostly administrative, and they have a profound impact on the way we live our lives. Chances are that building permits, zoning, road decisions, and even the boundaries of small cities and towns are controlled at the county level (Figure 12.2). From sidewalk widths to highway rights-of-way, the county may be your most directly responsible local agency when it comes to the health of your built environment. Your county government may have any number of officers and boards and also a county commission or council, and all of them make decisions. Some small rural counties may have only a judge and a sheriff.

Public health is what public health agencies do to improve the health of communities, but public health is also what a city or county planning department does when it designs communities in ways that will improve the health of the residents in the neighborhoods. Public health is what the park department does when it builds parks where we can relax and get physical activity and exercise. It is what's happening when a city creates a space suitable for older people and little children and everyone in between. Public health is what a transportation agency does when it installs public transit to help clean up our air, to get us safely and comfortably from one place to another, and to reduce the stress in our lives so we have time to spend with our families and our children. Public health is what the schools do when they educate our children. We know that educating children is the most important thing for their health, helping them to enjoy long, happy, productive, and healthy lives. We need to take a new look at how we plan communities, and build compact, multi-income, multiuse neighborhoods associated with quality public transportation. That's the kind of approach that we can do through planning, zoning, and good public policy to help make our environment healthier.

The State Government

States as organizations are big enough to have astonishingly large budgets for addressing issues in a substantial way, if they choose to do so. When I worked for the State of California, a state of thirty-five million people, the budget for the medical side of the state government was about $40 billion. Within that, my budget for public health was just under $300 million. That sounds like a lot of money, and it is, but it is a small percentage of what a state government spends on medical costs, and certainly a tiny fraction of what it spends on infrastructure and regulation of development that is producing the current lifestyles that are killing people.

People who work in state government can have amazing influence. Ed Schock explains that in order to have development, government needs to prime the location. "[Cleaning up pollution] is one of the roles that I think [local and state] government can play best. This isn't something that private developers can do—in fact they shy away from it. They say, 'Oh, if I have to get involved with environmental cleanup I don't want anything to do with [development at that site].'"

The Federal Government

The federal government gets a lot of headlines because when it acts, it acts in a big way. It controls billions of dollars for urban renewal, health, and regulation. When it comes to moving mountains, there's never been an organization like the U.S. federal government. That can be good or it can be bad, and I've seen it happen both ways. On the one hand you've got initiatives that the president can endorse, and suddenly you've got businesses and nonprofits and lower levels of government on board, but on the other hand the federal government can take forever to act and can become mired in political games. The federal government can, if it chooses to, fund organizations that do a tremendous amount of good. I've seen it. When I first came to work for the government as a young doctor, the organization I went to work for was called the Centers for Disease

Control. Now it is more than that. It is the Centers for Disease Control and Prevention because we've realized we cannot just control disease. We have to move upstream; we have to intervene and take action in the built, behavioral, social, and food environments that we live in.

As I mentioned before, the federal government has a wide variety of agencies that regulate everything from railroads to the amount of chlorine in drinking water. Your challenge is to find the agency that controls the issue you're working on, and engage its staff in putting agency resources and expertise behind you.

Nonprofits and Nongovernmental Organizations

Is there a nonprofit in your area that addresses your issue? Nonprofits exist to help in precisely the kind of process that you are beginning, and they've done it before, so they are a very useful resource. If you cannot find a nonprofit in your geographical area dedicated to your issue, find one in another area, and its people will likely point you in the right direction.

In the world of civic activism, the frontline troops are usually found in nonprofit, nongovernmental organizations (NGOs). These people opt for this work because it offers them a chance to focus on the specific issues they care about. What they can bring to your community and your project is experience in dealing with government and with business interests that you may lack.

Charleston's mayor Joe Riley has created nonprofits to improve his city. Charleston has received the Presidential Design Award and four HUD Blue Ribbon Awards for Best Practices, making it one of America's leaders in the creation of sustainable affordable housing. Mayor Riley fostered the creation of nonprofits dedicated to the creation and rehabilitation of affordable housing. He established the local Mayor's Council on Homelessness and Affordable Housing, with the goal of creating affordable housing opportunities in Charleston for very low income families. These organizations were instrumental in establishing a statewide housing trust fund and are working locally to develop a recurring source of funding for the development of affordable housing.

One of the key advantages of nonprofits is that they are used to working across boundaries between governments, businesses, and communities. Consider what Larry Cohen, at the Prevention Institute in Oakland, California, has created:

We've developed a tool called the *collaboration multiplier*, to look at the different sectors that are involved in a strategy or on a particular topic. Looking at these different sectors, what do they all do together or what are ways they can be engaged in an important topic to create a win-win? An example is agriculture—how do we get an agricultural policy in our nation? How do we make agricultural zoning decisions or zoning decisions about where fast-food outlets are or are not? How do we make decisions about menu labeling, which concerns business as well as health? How do we make those decisions in ways that advance health and solve multiple problems at the same time? How do we focus on community good rather than on the profit of a very narrow number of vested industries that make profits by selling something that increases the likelihood of diabetes, or whether they want to sell a drug after people get diabetes? Our tool provides a broader way of thinking.

Sometimes a good idea takes some time to develop to its full potential. Daniel McCarthy, who owns and operates the wine store in Detroit's Eastern Market, offers an example:

> The collaborative Detroit Ag Network [Agriculture Network], which is made up of a number of different nonprofits, began taking the surplus from its community gardens and selling it at the market to benefit those community gardeners. Two years ago [Detroit Ag members] were here for about ten weeks. Last year they were here for about twenty weeks. This year they're projecting a run of about thirty weeks because they have increased production to be able to keep the booth stocked. The booth sells food grown in Detroit and the proceeds go right back to the gardeners. It's one of the most popular booths here at the market.

> Farmers are getting their crops sold and the community is getting fresh, local produce. It sounds to

me as though this arrangement is solving two issues simultaneously.

Organizations, including foundations, have a clear purpose, and your issue may be just the vehicle they need to further their vision. Partnering with a nonprofit can provide funding for studies, lend credibility and expertise to navigate processes, and move an idea forward. Think of nonprofit foundations not only as funding sources but also as intellectual partners for improving the community.

Businesses and Professionals

Does your issue affect local businesses and professionals in your community? Could the problem be solved with the resources and expertise of these businesses and professionals? If the problem negatively affects business or business leaders, you may find ready allies in that community. Even if the issue is not directly related to business, many businesses will support change that benefits the community. If the issue can be solved with the help of a business or expert in the field, choosing one that is local can build further support for ongoing efforts.

Remember Tom Gougeon, a developer of Belmar in Colorado? As we mentioned in Chapter Four, he says, "Development is still a market-based business, but there has been a big change in the orientation of projects and in what people want. Real estate development grew into a bunch of specialists. There are people who did housing, people who did offices and those who did retail. Nobody put those pieces together." But in the last twenty to thirty years there has been a rediscovery in this business and in the professions of design and planning. "Belmar is an example of what can work when the professions of design and planning and of banking and finance work together to make walkable, livable places again." (Plate 2 is one example of downtown Belmar's mixed-use design.)

If you're looking to redevelop an area, a coalition with a financing package is needed. Securing funding is about leveraging commitments. Few like to be the first to sign on or to be part of a large account—especially where investment is concerned, most businesses want to say that they made the critical contribution. Consider breaking a large project into phases, so each piece is manageable for a funder to acquire, and that way a premiere funder can be highlighted for each phase. In-kind contributions of expertise and materials have a value that can be communicated as leveraged resources in conversations with financial backers. Realize that a phased project can open opportunities for existing funders to participate again later.

Brian Leary, former AIG Global Real Estate vice president of design and development for Atlantic Station, is a business manager who used his business to change the built environment. He talks about the stakeholders necessary to change an environment and how success builds on success. "We had one good idea. Let's take a steel mill and clean it up. We've had a lot of great ideas along the way, and we've wanted Atlantic Station to be a resource,

not only for Atlanta and Georgia but for the entire country about how to do things differently. We keep our doors open for developers, mayors, council members, EPA [the Environmental Protection Agency], and researchers to collect data and share what we've learned. Our goal is to not only do it better going forward but to export the idea across the country."

And Will Fleissig, of Continuum Partners in Colorado, says, "I've always believed that in the end, what people care about are other people. People want to be around others and they want to be in places where they can be with and watch others. This is why urban places work. It's the human dimension of the city. So for us, developing is the art of creating the physical environment in which people interact."

Builders and architects have a special responsibility when it comes to whether a building that is being created will be an unhealthy or a healthy structure. By the decisions they make each day, they create and guide development. As businesspeople they need to deliver the buildings their customers demand, but they have tremendous ability to guide the customer and offer solutions that are in the long-term interest of both the customer and the community at large.

In a number of the places we've visited in this book—in Belmar, Charleston, and Detroit, for example—businesses have been a part of the built environment and public health problem and also a part of the solution. It is usually just as profitable to be part of the solution, and owners and managers of business are often part of an affected community themselves. If they do not see their relationship to your issue, ask them why not.

Academic Institutions

Are there academic resources that could help you? Like nonprofits, academic institutions exist in part to research the issues that drive people. Academics may be interested in your project even though they do not live or work near your community. Contact the nearest university with a department of architecture, public planning, or public health. Even if people there cannot help you, they may know of others in the academic realm who are working on ideas related to your project.

Universities and colleges employ professors, researchers, writers, and scientists to do far more than teach classes to current students. They have access to an almost unlimited group of students who can help with design and implementation and who can also analyze information as part of a school project. Consider partnering with a professor or graduate student to facilitate a community assessment, implement and analyze a survey, compile an exhaustive bibliography of research on a topic, or compile findings on your issue or the work done to date in your area. By providing a real-world context for students to learn from, you are providing a much needed service.

I offer a course on the built environment and public health, and the students are hungry for case studies that integrate multiple issues, but our academic institutions are *stovepiped* into isolated topic areas. Students arrive saying, "I am worried about the future of the planet, I am worried about energy efficiency, I am worried about my children, and I am worried about paying my bills."

We do not have enough courses that integrate public health and the built environment. We should integrate energy efficiency with engineering at a very high level, and sustainability with architecture. We have got to reinvent academic practice to deal with twenty-first-century problems, and having people who stay isolated within their areas of expertise isn't going to help. If we can ask the stream of young people going into advanced education to participate in case studies that reflect the real world, we would actually make learning about public health more exciting and cross-disciplinary. If we had projects that forced departments to work together, we could actually make things happen. When students see that they can effect change, they get excited about being in public health.

With the collaboration of a professor who is also a researcher and a scientist, you will also have an advocate for realizing next steps. Academics have prestige that can be leveraged to convince government, business, and the community that a particular change needs to happen. Through publications, the human plight associated with your issue can be communicated to a global audience, gaining newsworthiness and resources. This win-win perspective is how you build coalitions—with people taking on your work and making it more than what one person can accomplish.

Depending on the issue or concern, the K–12 setting can be a great resource as well. Through service-learning projects, students around the country are learning their academic subjects through solving real-world problems in the community. If your project connects with a curriculum, or if you think it might, contact the local school to see whether a teacher or group of students is interested in working with you. Students can be a huge asset in gathering basic information and getting the word out about a new initiative. In Elgin, Illinois, and across the nation, students do water quality testing, help to clean up open spaces, remove nonnative species, implement recycling efforts, and participate in forums on community sustainability. Imagine how powerful these young people will be as adults. This is their community—they have created it in their image. Their ownership and pride in the improvements they have helped to make is easy to see across their faces.

Faith-Based Organizations

Churches have credibility within many communities and can be employed as allies for change. Can you enlist one or more faith-based organizations in your area as an ally? When it comes to human needs, faith-based organizations are often ready to help, and their leaders may have tremendous credibility throughout the community.

Having an effect on a community requires enlisting the institutions that people trust. Faith-based organizations provide service across generations and are often part of the fabric of family life. A church, temple, or mosque is more than a place for worship—it is a place that understands a call to community and community action. It is often the center of community activity and an institution that has been caring for the elderly, infirm, and poor for decades. Architect Roland Wiley defines a sustainable community as one that has "five spokes to the wheel. A sustainable community is a place where you live, learn, work, pray, and play." When faith-based organizations see your issue as furthering their mission, they will understand and support your efforts to improve the built environment.

Recall Joseph Williams, the CEO of New Creations Community Outreach in Detroit, Michigan, whom we met in Chapter Ten. NCCO is an example of a faith-based group that works with a number of churches in order to provide comprehensive assistance to its clients, individuals who are in transition from prison to the community, and to work to break the cycle of poverty that can lead to failures of education and to incarceration.

Faith-based organizations are in the business of building community. One of the greatest strengths of faith-based organizations is that they know how to spread the word about events and issues. Donele Wilkins of Detroiters Working for Environmental Justice remembers when the United Church of Christ Commission for Racial Justice completed its report on race and toxic waste in America, and convened a group to share the results. "They thought maybe 300 people would show up to hear the results, but over a thousand came, hearing about the event by word of mouth. They just showed up and flooded the hotel."

Citizens and the Global Community

In the end, the built environment must serve the community of people who live within it. If it does not serve them well and an effort must be made to change it, sooner or later the work of that effort comes down to the community. A community that is unified in its purpose can direct government action, provide strength to nonprofits and nongovernmental organizations, bring along the cooperation of businesses and professionals, offer academics a practical application for their work, and ultimately make a new environment work.

"If you do not have the community involved, how will the work get done which could be years in the making? Policy changes don't happen overnight, but that doesn't mean we should sit and twiddle our thumbs. We need to get going and move all the wheels at the same time," urges Nsedu Witherspoon.

SOCIAL NETWORKING

It is easier to effect positive change now than it has ever been in the past, in large part because it is easier to communicate with others instantaneously. Social networking tools such as Twitter, Facebook, citizen journalist sites, e-mail lists, and blogs can help you to gather, organize, and communicate with the people you need to reach to move your project forward.

The key to social media is to get the communication channels established early and then use them regularly, with high-value content. Use these guidelines as you work to harness the power of social media:

Build. Create social media channels early, so you can share the Web addresses with every person you meet and print them on every flyer and every sign you make. Maximize people's ability to find your group and information on your issue.

Cross-pollinate. Post ideas on various platforms to maximize coverage. Make sure people know you're out there by posting information and links to your social media channels on related Web sites, blogs, and e-mail lists.

Delegate. Someone, just one person, must *own* the communication channel and be responsible for scheduled dispatches. If the job belongs to everyone then it belongs to no one, and it won't get done.

Offer high-quality content. Make sure all communication is of real value to readers, or they will quickly jump away to other sites. If you bore readers with multiple messages each day and useless filler, they'll unsubscribe or stop visiting. Send information with important updates, not useless cheerleading.

Visualize. Relevant photos and graphics enhance your communications and draw attention to the nearby text.

Designate. Calling for volunteers through a group channel won't meet most of your needs for help. If you need someone to do a job, ask that person individually.

If you can harness the tools available for free on the Internet, you can spread the word, offer your community a chance to participate, and prove the popularity of your program to skeptical stakeholders. But social media are not the answer for all your communication needs. Traditional means of organizing and advocating and efforts to make personal contacts are still critical as well. Moreover, as Nsedu Witherspoon points out, "Not everyone is on the Internet. There are still communities that are very disconnected for economic reasons and access issues. It's not enough to say, 'Just go to our Web site and download it.' That's just not realistic, especially for high-risk communities." One way to address this problem when you are giving out Internet links is to include the location of free Internet access sites, such as public libraries.

GETTING EVERYONE TO PULL TOGETHER

Ryan Gravel is the visionary behind Atlanta's Beltline and an urban planner at Perkins+Will. The Beltline is a redevelopment of a largely abandoned railroad right-of-way circling Atlanta. It was Gravel who first proposed putting new transit there; now the Beltline is becoming home not only to new public transportation but also to parks and trails, mixed-use development, and affordable housing. It is now the largest redevelopment project in the United States. (Plate 23 shows a section of the Beltline when it was under construction.) Gravel understands the

importance—the necessity—of getting the community together and, as much as possible, invested in the process of rethinking the design of the built environment:

One of the reasons for the Atlanta Beltline is it's a way to transform the city through a grassroots community movement where you get community buy-in. The Atlanta Beltline began as one person's idea. Before long there was a health impact assessment, funded by the Robert Wood Johnson Foundation. Identifying a quantifiable health issue and providing a potential solution built momentum that brought others on board. This project included academia, nonprofits, businesses, and a variety of government agencies. If the Beltline had come from the mayor's office as, "This is our plan. We're going to implement it now," then I don't think we would have gotten as far as we did. It was because we spent three years attending hundreds of neighborhood meetings, building that buy-in, that we were able to create enough momentum to get the attention of the elected officials and the regional planners to move the project forward. In the end, the Beltline project belongs to all of Atlanta, not just one stakeholder group.

So think about your issue and who might have a stake in addressing it. Do a bit of research on the topic and find out who has jurisdiction. Once you have a list of potential stakeholders identified—just identified—then you are ready to bring those stakeholders together to set goals and create an action plan.

Chapter 13

Create an Action Plan

Worrying about a problem and *talking* about a problem are not the same as *solving* a problem. Solutions nearly always require action, but determining the *right* action takes knowledge and planning.

If you were a physician and your patient—in this case, your community—came to you for help, you would analyze what the patient communicates, examine visible signs and symptoms carefully, and often order some tests. It is essential to *make the diagnosis*. It can be disastrous for a doctor to treat based on one disease when the patient has a different illness. For example, if the doctor thinks the patient has an autoimmune disease and prescribes immunosuppressant drugs like steroids when the patient really has HIV or TB, the steroids will reduce immunity and worsen the illness.

Making the diagnosis and starting treatment are important, but they are not enough. A good doctor must continue to *be present* for the patient. Sometimes the treatment is wrong, or it does not work as expected. It is a poor doctor who abandons a patient when the patient feels helpless. Apothecaries, the pharmacists of yore,

kept in their cabinets the herbal medicament *tincture of thyme*. When I was in training, senior physicians would occasionally offer, in a play on words, that a *tincture of time* was the only workable treatment. The message was always clear: only time, not medication, would bring healing. In the same vein, the word *patient* means one who *endures with forbearance*. Often communities are patients, as time is needed to heal some of their wounds and illnesses.

ANALYZE THE SYMPTOMS

You, as the doctor to your community, have many assets as you move into this phase—you have read about the built environment and ways to make it healthier; you have completed an initial analysis of your built environment, which helped you identify what the issue to be treated might be (see Chapter Eleven); and you have identified a group of people whom you can bring together to review your process and findings (see Chapter Twelve).

Many health impacts resulting from our built environment disproportionately affect vulnerable groups. To create a community that works for everyone, we must consider the needs of these multiple populations and we must avoid developing economic and social monocultures. In designing and redeveloping your built environment so that it fosters independence and economic, social, and ethnic diversity, what would you change?

Universal design is an approach that says products should be designed to be usable by everyone and shouldn't have to be adapted for people with special needs. If we were to apply the principles of universal design to our homes, the living space would be on one floor (even when the building is multistory), with smooth entrances and floors so strollers and wheelchairs can enter and navigate the residence. Although it is good for our bodies to be lifting things and climbing stairs when we are young, at some point in our lives, due to aging or illness, these activities can be difficult and tiring. So, in our well-designed homes, doorways would be a little wider than the current standard, and we would have lever handles, which are much easier to use, especially if we are carrying packages or a baby or if we have arthritis.

Our cupboard shelves would be on sliders, so we would not have to kneel down and dig to the back, lifting heavy items to find what we need. A faucet would be placed near the stove so heavy pots would not need to be carried across the kitchen to be filled. My mother, who is in her eighties, complains about kitchen counters that are too high and make it hard for her to cook. Our homes might have a space under the sink that would allow us to slide a chair or wheelchair under the counter and sit while doing dishes or washing food. Our homes would also have at least two bedrooms and two bathrooms so that we could comfortably accommodate an in-home care provider during serious illnesses or when we need extra help late in our lives.

By designing our homes in these and other ways to last a lifetime, we would not only be able to remain in our communities and to make our lives more comfortable but we would also reduce the need for elder care facilities, bringing down health care costs. When we design a home to nurture us, often the little things make the biggest differences.

Complete Streets

Roads are important and enduring parts of the built environment that influence our health. When we build roads only for vehicles and we do not build them for pedestrians, bicycle riders, and people who need to go about their life's work without driving, the roads threaten our health. Roads designed to meet the needs of all members of the community are known as *complete streets* and include at least sidewalks, crosswalks, drainage, and lighting.

Healthy Buildings

Design matters when we build. Our health is shaped by design choices, often made long before we arrived on the scene. The American Institute of Architects (AIA) has issued a call to action to create carbon-neutral buildings by the year 2030.[1] Buildings, from homes to office towers, make up the largest single contributor to the production of greenhouse gases in America, largely because they need to be heated, cooled, and supplied with lighting and elevators and because so many buildings designed in the past need to use large amounts of fossil fuels to make up for their inadequate insulation, single-pane windows, inefficient motors, and inadequate ductwork. We should respond effectively to the needs of both natural and human systems.

When we built sealed-up, well-insulated homes in the late 1970s, they did not work as well as intended because gases, molds, and moisture become trapped inside, and many people suffered the burning eyes and throats known as *sick building syndrome*. These hazards can be remediated by installing conduits for negative airflow, to suction stale air found in bathrooms, kitchens, and basements out of the house, and conduits for positive airflow, to bring temperature-balanced fresh air into the living spaces. Energy savings and efficiency can be maintained by using highly efficient heat exchangers.

We need appropriate amounts of light. Inadequate indoor light, including insufficient daylight, presents safety risks and makes reading, studying, and socializing more challenging. Too much darkness during the day is linked to depression in some individuals (including myself); too little darkness at night can impair rest.

Buildings with inadequate exits and fire resistance present safety risks, as do buildings with inadequately reinforced foundations, roofs, windows, and walls, especially during storms and earthquakes. Buildings with ugly, narrow, inaccessible, and sometimes locked stairways discourage stair use.

Single, unattached, or one-story, poorly designed, built, and insulated structures are less energy efficient than a dense set of well-built structures. When structures are spaced out so that people have to travel greater distances, transportation and energy costs and pollution are increased. As more capital is directed to pay for inefficient use of space in the built environment, less is available for other beneficial uses, such as education, health, mass transportation, paying the mortgage, or building financial equity.

Public Transit and Safer Shipping

When we look through the lens of social justice, we see that it is essential for communities to invest in public transit that is clean and affordable and can be used by all citizens. Heavy diesel trucks do not belong in neighborhoods and should not use residential streets as shortcuts between ports and distribution centers, as is often done in Oakland and Los Angeles. The Port of

Los Angeles claims to generate over $402 million in annual revenue,[2] yet the port claims it does not have the funding to deploy a high-quality, preferably electric, rail system to move containers to distribution sites fifty miles inland. According to the national Fatality Analysis Reporting System, 7.1 percent of all vehicles involved in fatal crashes in 2009 were heavy trucks.[3] Heavy trucks damage road surfaces; produce substantial air pollution, especially in the form of particulates; and are much noisier than automobiles and light trucks. Reducing the number of trucks on the road would reduce these problems and would also benefit communities by increasing the demand for local production of food, goods, and services, as well as waste transfer.

There are changes under way. As reported in Chapter Nine, for example, the Port of Oakland is no longer using diesel-powered cranes (Figure 13.1).

Neighborhood Design

Smart redevelopment can reduce urban blight and create affordable housing and parks, schools, playgrounds, and other public facilities. For example, Octavia Boulevard in San Francisco was once part of the city's two-level Central Freeway, sitting alongside shadowy, fenced-off land before it was damaged in the 1989 Loma Prieta earthquake. Following the earthquake, Octavia Boulevard was redeveloped and redesigned to better use the freeway's right-of-way for additional street space and new housing. Today, quiet outer roadways leading to and from the rebuilt freeway are lined with homes and businesses, connecting faster traffic with the inner roadways. A new park has been created as part of the boulevard project, and in recent years the local neighborhood has experienced a boom in housing and business development.[4]

Children who lead sedentary lives are at risk for a variety of health problems. Parents cite traffic danger as the major reason why their children are unable to bicycle

Figure 13.1 Installing electric cranes and reducing diesel truck emissions has improved the health of Port of Oakland employees.

Source: Photograph from the Media Policy Center.

or walk to school. The nationwide decline in walking and bicycling to school has also greatly increased automobile traffic congestion and air pollution around schools, which further degrades pedestrian and bicycle safety. The purpose of the federal Safe Routes to School program (http://safety.fhwa.dot.gov/saferoutes/) is to address these issues. The program supplies funding and staffing to empower communities to make walking and bicycling to school a safe and routine activity once again.[5]

Parks and green spaces are essential pieces of any healthy environment. The American Society of Landscape Architects (ASLA) advocates appropriate use of vegetation in the built environment because it has a significant influence on the quality of human life and can reduce energy needs. Trees and other plants filter pollutants from the air and water, mitigate wind, reduce solar heat gain, stabilize soil to prevent or reduce erosion, create animal habitats, help filter and absorb stormwater runoff, and may help control carbon emissions.[6] Furthermore, parks provide communities with opportunities for recreation, increasing social and financial capital.

As I pointed out in Chapter Three, Richard Florida, author of *The Rise of the Creative Class*, reports that *cultural liveliness* is a measure that has been used in ranking the best American cities. For example, a high prevalence of small local restaurants was seen as an asset compared to an abundance of franchise eateries (Figure 13.2; also see Plate 25). In support of this idea, Florida emphasizes the "need for a vibrant cultural scene to attract the most creative and productive workers to an

Figure 13.2 Small local restaurants are often healthy options and can foster a dynamic and active local social life.
Source: Photograph from Media Policy Center.

area—especially young people. Cities with lively cultural opportunities tend to generate more wealth during good times, and can better resist economic downturns."[7] We are, after all, social animals, and our choices in how we build our communities can and should provide us with opportunities for a dynamic and active local social life, rather than unhealthy options that benefit only the bottom lines of various corporations.

Business Environments

We enjoy the experience of walking through a store, picking up items on sale, testing and feeling them, or tasting new foods. Restaurants and shops that are locally owned and not franchised give a unique character to a community and are an important part of a lively neighborhood. Our built environment is not only policy and culture in concrete but also a framework for socializing.

The built environment supports our interaction with all types of businesses when it makes them easily accessible by everyone. Inadequate public transportation and parking that is too far from shopping establishments, too expensive, or too dangerous discourages casual shoppers. Long commutes and the search for parking will dull any consumer's shopping enthusiasm.

Pollution

The wealthy do not often live in polluted places. They can afford not to. It is usually those with the fewest resources who have to overcome the greatest challenges, and such challenges are inherent in a poorly designed built environment. Angela Reyes, executive director of the Detroit Hispanic Development Corporation, is working to change Detroit, but it is an uphill battle and pollution is a major factor in that battle. Reyes says, "The built environment is only one of many serious issues facing the city. We have heavy truck traffic from the Ambassador Bridge, incinerators spewing soot near schools, polluting industrial sites, and people dumping in the middle of the community. It is very difficult to know where to start. We have a lot of people with tenacity who aren't willing to give up, but the issues become overwhelming."

Social Equity

"Cities that lose residents to the suburbs suffer from a decline in business investment and tax revenues, yet still must pay for a large physical infrastructure and the disproportionate social problems (crime, drug abuse, welfare) of those who cannot or do not leave," reports Smart Growth America.[8] Investing resources in existing neighborhoods, and encouraging affordable housing can create more equitable communities. Mayor Joe Riley's comment about low-income housing that I reported in Chapter Six is worth repeating here: "We should not accept anything less than beautiful, properly designed . . . housing for low-income people. It is not acceptable to allow the [government] to build that disgusting, auto-oriented Department of Housing and Urban Development (HUD) public housing. . . . We can refurbish old Victorian houses to provide affordable housing. It not only works well for the neighborhood, but is gorgeous as well."[9]

Urban disinvestment is the natural outcome of deliberate policy choices. For example, federal funding of highways reduces the focus on mass transit. Local exclusionary zoning regulations result in barriers to creating affordable housing and repurposing vacant properties.

DETERMINE THE DIAGNOSIS

It is time to assemble your stakeholders and look at the information you have gathered. Each stakeholder will bring a different perspective, set of information, and expertise to the process. Get to the root causes of poor health in your community. If the problem is widespread obesity, is it caused by the lack of locally available healthy foods, lack of opportunities for people to engage in physical activity, or something else?

Making a diagnosis means sorting out the signs, symptoms, and data to identify what is really going on. When the diagnoses are right, smart solutions can be identified; often most or all of the symptoms will come from the same root cause. The ideal treatment is one that remedies many ills and keeps them from recurring.

IMPLEMENT THE CURE

To make significant change in your community, you need to know where you are headed and then get started. Although the work may not move straight ahead all the time, there are four general steps to success: utilize, evangelize, organize, and realize. Each step takes you and your stakeholders closer to accomplishing the defined goals.

Utilize

Learn from what has been done before and follow relevant examples. Your situation may be unique in some ways, but others have encountered similar situations, and you can glean valuable skills and insights to apply to your situation.

Get Inspired

Be influenced by someone, and use that person's life and work as a model for what you would like to do. There are amazing leaders all around us. Some inspire by what they write or say; others by the choices made during challenging times. To be inspired by someone else is not to be doing exactly the same things that person did but to take lessons from that leader and apply them to your life.

Imitate

Another way to utilize what others have done is to examine their actions carefully. How a person communicates with and modifies a message for various audiences is what the art of public speaking is about. As a pediatrician, for example, I have a way of talking to doctors that other doctors can understand. When I talk to elected officials, I have to speak differently. Politicians do not have the technical background, the time, or the need for all the data. They love and need stories and "sound bites" that they can use later. Mainly they need a simple, clear message they can understand, can communicate to others, and can act on. So, I use different words with them, but I still have the technical information available should they or their staff want the data.

Apply Aggressive Failure Analysis

When there are hard problems to be solved, we should try a bunch of strategies to figure out what will work. We also have to be OK with some things not working. We have to learn valuable lessons from our failures. Airline safety agencies and people in the airline business call this *aggressive failure analysis*. They examine failures in minute detail to determine tools for improvement. This is why commercial air travel is so remarkably safe today.

The fog that descended on London in December 1952 was what was known as a *killer fog*. The black, acrid, dirty, warm smoke and fog that descended on the city was so thick that movies and theaters were closed—the audience could not see the stage. It was attributed to the burning of coal for fuel, which created sulfur oxides in the cool, damp winter air. This historic event caused an astounding increase in premature morbidity and death in Greater London (Figure 13.3). Such extreme events are indicators of what can happen under circumstances of meteorological inversion and concentrated

pollution. It took 12,000 excess deaths for people to take notice of the smog they were creating and realize that something had to change.[10]

In 1943, the smog in Los Angeles was persistent, with climbing rates of chronic lung damage, burning eyes, and chronic headaches. As the city's population had expanded, the number of cars and trucks on the road had gone up in tandem. This brown smog was caused by automobile exhaust, which formed nitrogen dioxide and ozone. When the warm land and cool ocean caused a temperature inversion, these chemicals were trapped in the Los Angeles basin and accumulated. Through the work of researchers, regulators and industry, the source of the smog was determined, and new air quality standards, the most stringent in the nation, were put in place.

Figure 13.3 Approximate weekly mortality and SO$_2$ concentrations for Greater London, 1952–1953.

Source: M. L. Bell and D. L. Davis, "Reassessment of the Lethal London Fog of 1952: Novel Indicators of Acute and Chronic Consequences of Acute Exposure to Air Pollution," *Environmental Health Perspectives* (2001), *109* (Suppl. 3), 389–394.

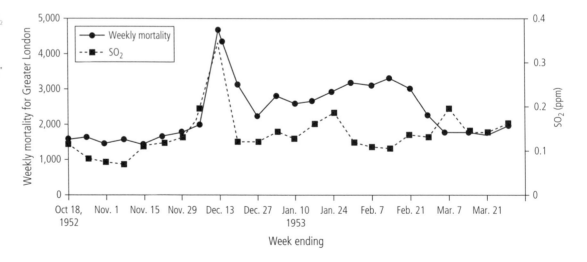

Car tailpipe, industry, and commercial emissions were regulated.[11]

There is an inherent opportunity in disasters. There are numerous reasons why we give insufficient attention to infrastructure until it fails, but when it does fail, do we know where to expend the resources given to address the problem?

The Chinese economy has been growing and is now the world's second largest economy. As the Chinese population and individual personal wealth grow, people are buying conveniences like cars. Car sales in China increased by 45 percent in 2009.[12] Food is grown in the countryside and business is conducted in the city. Trucks bring food and goods into the city to meet the growing demand. The Chinese government has a lot of smart people, so in August 2010, was anyone really surprised that a combination of sandstorms, construction, and industrial projects resulted in a traffic jam along the Beijing-Tibet Expressway stretching an incredible sixty-two miles and lasting nearly a week[13] (Figure 13.4)? This could have been predicted through simple modeling and logic. The Chinese government is now making immense investments in rail infrastructure, including high-speed rail. Between 2008 and early 2011, China increased its high-speed rail capacity from 849 km to 8,400 km and is targeting 19,000 km by 2014. This investment is the greatest public works investment since the United States built the interstate highway system in the twentieth century.[14]

You can learn from failed endeavors and from poorly planned strategies. There are lessons in unsuccessful

Figure 13.4 The worst traffic jam in the history of the world happened in China in August of 2010.
Source: Photograph from the Media Policy Center.

efforts at change that you might employ in conversations. Sometimes the appropriate technology was not ready when a previous effort failed, and now it is ready. Someone else's idea, implemented poorly, could be successful in your hands.

Find What Works

For former Elgin mayor Ed Schock, success is measured by whether his ideas are ultimately imitated. "I think Elgin has an opportunity to do something incredibly special and valuable. If we succeed we will be able to demonstrate what we have done to others. How can Elgin be sustainable if the others around us aren't? What have we really accomplished? Very little, I suppose. Our effort has to be viewed in a larger context, which is, if we can provide an example, encouragement, or motivation to other communities, so they undertake and replicate the same effort, then we will have really accomplished something."

Another way to utilize is to look at successful projects and use elements from them in your project. As Ryan Gravel, urban designer at Perkins+Will, says, "Atlanta is the poster child of sprawl in America. I think that as we start to find solutions to the problems created by sprawl, like the Beltline, there will be a lot for other cities to learn from it. I think the lesson from the Beltline is not that every city has to have a circle of railroads around it but that we can use strategies about how the design of infrastructure can affect multiple public benefits like public health or economic development."

For some, it's about being on the cutting edge of innovation rather than replicating existing work. Tom Gougeon, of Continuum Partners in Colorado, has a mission to change the world. "We can change the world not by doing something a thousand times but by doing something once or twice that is hopefully better and different. We hope the rest of the world looks at what we did and learns from it," he explains. "Our goal is not to do Belmar again ten times but to figure out what's next because that's where our opportunities really are. We want to keep working on the same urban principles of diversity and walkability, but create smarter, greener, more humane and groundbreaking urban settings. We learn by doing trial and error because you can't go study

the answer and then carry it out. The answers don't exist yet."

Evangelize

Evangelizing is about getting your ideas out into the world. It's about putting beliefs and ideas into a context that even those with a different worldview can relate to. "You have to believe in the righteousness of your cause, and if you've been careful and evidence-based, you have to stand on your conclusions," says Chris Kochtitzky of the Centers for Disease Control and Prevention (CDC).

Evangelism involves telling a compelling story, being a good leader, having credibility, getting out the message, keeping control of the situation, and taking bold steps.

Be a Storyteller

When I became a physician I had to be very analytical. The training was all about the facts, not about the story. Every patient has a story, but we had to find statistics. About five or ten years into my public health experience, I began to reintegrate myself—the physician and the storyteller. Letting that storytelling person out was far more interesting than being the data-driven person. We cram people with information, not knowledge, and they cannot process. Something has to give. When you want someone to see the world as you do, giving him or her more information is not necessarily what is needed.

Maybe an analogy or a story will answer the questions and disarm the listener at the same time.

Be Aware That Leadership Matters

If someone had told me twenty-five years ago that leaders matter, I would not have believed it, but it is true. The whole federal government takes on the personality of the current president. When the directors at the CDC change, it affects an organization of about 14,000 people. You could be the smartest person around, but it's your fundamental humanity that has influence throughout the system.

In Charleston, Mayor Riley kept the developers from turning the waterfront into a private access area. He made a public space instead. People are smiling because of Joe Riley. The fountains, parks, and view of the water—this is priceless. If Joe Riley had not been there, what would Charleston have become?

A lot of people would view *walkability* as a nice amenity rather than an essential prerogative. I think that walkability is a central issue in creating places that work for people. Being able to go outside and walk is a human need. This is as essential as oxygen and food. If you cannot go for an interesting walk unless you first walk two miles from where you live just to get to something worth seeing, that is bad. It comes down to profoundly failed leadership.

Look at the Embarcadero in San Francisco. There were transportation plans to build an elevated highway along the waterfront of San Francisco about 400 feet from the shoreline, but in 1959, mothers with baby carriages went out and protested. The *great freeway revolt* caused the San Francisco Board of Supervisors to acknowledge the opposition to certain freeway routes and change their plan. The whole episode left the community with a distinctly antifreeway sensibility. This was leadership. Now you can walk along the Embarcadero and everyone's smiling. There are people walking, street performers, joggers, bicyclists—this place went from a dirty, scary, unpleasant place to a place of beauty and health.

Recall that the World Health Organization defines health not just as the absence of disease but as "a state of complete physical, mental, and social well-being." How will your leadership increase the health of your community?

Build Credibility

This is about branding yourself and your team. You may be an expert yourself or you may be the point of contact that can find the right expert for the situation. Both roles are incredibly valuable. Here are a couple of examples from my own experience of being the outside expert.

Teri Duarte, health program manager at the Sacramento County Health Department, once asked me to make a presentation at a meeting on land use and public health in Sacramento, California. Although Teri Duarte has deep knowledge in her field, she knew that bringing in someone from the outside as a keynote speaker to jumpstart discussion can be valuable. Later she reported that "a local developer on the agenda is now holding a focus group on health for a proposed land-use project next week. He wouldn't have chosen this direction if he [hadn't heard Dick Jackson]. Dedicating a whole focus group to health is a huge impact on the development project."

Marice Ashe, executive director of Public Health Law & Policy, recalls how she and I worked together on a presentation at the CDC:

> He has credibility and he's an inspiring speaker. He talked about the drop in heart attack rates in the emergency rooms around the time of the Olympics in Atlanta, and I talked to him about zoning controls and other land-use tools that can be used to reduce exposures. He's the big-picture guy and my staff follows him to go over all the technical details of how a community can implement the ideas. It's a wonderful synergy. We complement each other and that makes for a more powerful presentation.

Decisions often happen around a very small table, so getting a seat at that table is paramount in becoming an agent for change. Often those who have the power to make changes in policy need information, so do your homework. Be the "go to" person or the group that others turn to when they want unbiased and accurate information on a topic.

Get Out the Message

The only way people are going to know what you're thinking about and what you're doing is to tell them.

There are many ways to "get the word out," from using the mass media to social networking and making personal connections. A combination of approaches may be needed, but the right combination will be based on the issues and the culture of your community.

The most important part of getting out the message is deciding who your spokesperson or people will be. As Anthony Iton reminds us, "your integrity and reputation are all you've got." Credibility is key. Al Gore did some wonderful things to raise awareness about global climate change, but remember what happened when a group looked at his personal carbon footprint? Being a spokesperson is more than a speaking engagement—this person must *walk the talk* in a way that is beyond doubt.

Getting your message out means you're engaging in politics. You do not have to run for office to put something on a ballot—influencing public decisions is still politics. Local politics is a retail business in that you have to contact and interact with people individually to achieve the goal. Politics is about pounding the pavement. This is a lot of work, but it makes a difference. When you start branching into traditional media markets—radio, television, and newspapers—making a credible first impression is vital. Put your best spokesperson forward, make clear, quotable comments, and you could become a regular top story, not just a one-time event. You could become the resident expert on the topic, informing the public on progress toward a goal the community cares about.

Sacramento, California, like a number of other communities across the country, is bringing about a downtown renaissance. With proper planning and communication across various heretofore unrelated disciplines, it is building the concept of a healthy community in concrete. Connecting the built environment with better public health is an idea that has gone viral. Planners like the integration of health into planning decisions, and now it's coming *from* them rather than happening *to* them. Planners and developers are integrating these concepts themselves. Sacramento's planning staff have just completed a food desert study. They are working on a health impact assessment (HIA) that will gauge the effect of the lack of good quality fresh fruits and vegetables on public health. Such preparation coupled with market forces favoring walkable, lively communities, has resulted in an explosion of restaurants and plans for dense walkable residences in the downtown area. There's a tidal wave of interest.

Often groups put together events to attract people and send a message (Figure 13.5). Earth Day concerts, Rock the Vote, and other music and media blitzes bring like-minded people together and hammer home a message of action, but be careful whom you invite. Everyone who presents must be *on message*, to keep your issue from being diluted.

An increasingly critical element of getting the word out is strategic use of the Internet and social media. E-mail, e-mail groups, Web sites, social networking sites, YouTube, and the like, all provide structure and opportunity for people to find you, just as you can find resources using the same tools. There is amazing power in reaching

Figure 13.5 Events like Earth Day are excellent opportunities to get your message out and increase your visibility.
Source: Photograph from the Media Policy Center.

a global audience, but with that power comes a responsibility not to waste people's time. Here are a few suggestions for you to consider:

- *Get your domain name early.* Put your Web site and e-mail addresses on all your print media.
- *Stay focused on your mission.* Use social media to communicate your issue and document your progress toward a solution. Add links that support your efforts and acknowledge sponsors, but keep your "brand," your issue, as the focus of the page. If you have other issues you're working on, create other Web sites for them and link to them, so your message does not get diluted.

- *Be memorable.* If all your communications have similar branding (color scheme, logo, look, and feel), then people will get to know your brand and come to rely on it. Build your cachet so that people will trust material because it came from you.
- *Capture contact information.* Build databases of contacts so you can send out information to interested parties in a timely fashion and make requests for information, services, and expertise when help is needed.
- *Be aware that what is posted never completely disappears.* Even when links to your Web site are gone, search engines that have captured and stored your postings will be making them available. So, reflect on how you want to be remembered, and make everything you put on a Web page count. This is a public forum and it lasts forever.

Keep Control of the Situation

As you engage others in your work it is easy to get overly inclusive of others or sidetracked by others. Both results have the same effect—they dilute your message. Even though you want to be a part of a growing coalition focused on the solution to your issue, not all endorsements are worth expending the political capital needed to get them.

Take Bold Steps

Your proposal, speech, logo, or blueprint will never be perfect, so put it out there anyway. You brought your stakeholders together for a reason. If they form a balanced group, you will have critical as well as friendly voices around your table. In the process of making change, these documents will go through revision.

When Chris Kochtitzky and I were at the CDC, we wrote our first article together on the built environment. It was part of a series of white papers, or monographs, for Sprawl Watch Clearinghouse. We were asked to look at the health implications of sprawl. We never intended our work to "tip off the world," as Chris would say. We did not make huge claims and the paper was well footnoted, but its timing was pivotal. Other interests were poised on both sides of the issue. On one side, developers and planners felt their worldview was being challenged by our ideas, that this was a threat to their livelihood. On the other side, people were hungry for a solution to planning challenges and the levers available were not working. There was a pent-up energy among people waiting for a new solution, so when we published, everyone looked.

What we did was to go back to what Frederick Law Olmsted believed. He designed Central Park in New York City in order to give people a "place for fitness," or what is now called physical activity. Our intent was to encourage people to rediscover lost knowledge, to return to a place before their communities chose the path they're now on, and then to rethink their direction.

At the same time our paper was published, the CDC published a set of raw data without interpretation. People who felt threatened by our ideas grabbed those data as evidence that we were wrong. Things got very heated, but the point is that we put our ideas out there and gave people something to react to. In the end, these same dissenters have used our work to plan and develop projects in new and smarter ways. They have just as much work as they always had, maybe even more, and healthier communities are emerging.

When you move an initiative or a solution forward, people are influenced "by the logic in your argument," says Tony Iton. When he was working on health issues related to the Port of Oakland in California, Iton challenged the right of the port to kill the people in the surrounding community for economic gain. It was a bold argument, but he had the statistics to make a reasonable assertion. As an advocate for public health, he became an adviser to the port commission and influenced the mayor in development decisions. This changed the posture of the port and expanded the capacity of public health in the area. As a result there has been new work on land use and design development in the area.

Organize

When you have a great idea, the hardest part can be anticipating and planning how to get it off the ground and implemented. Organization is important. You must have a logical plan that is simple and flexible and that uses the resources available. Now that you have defined your goals, list specific issues related to those goals,

analyze them as a system, develop indicators, and select comparison standards. Create a profile of each issue, and rank the issues in order to set priorities. Develop a plan to implement each priority, and put structures in place to evaluate progress and effectiveness of each. Give each stakeholder a specific focus, and check in with each stakeholder frequently. Include times to come together and refocus; as the environment changes, so must the plan.

West Wabasso is a community located in Indian River County, Florida. About five years ago, I met with public health officer Julian Price to help a group of community members figure out how to improve their community. They decided to follow PACE-EH, an iterative process that cycles as many times as needed through assessing the community, defining the problem, taking action, and evaluating the result. At the time, Wabasso was about as close to a third-world community as anywhere in America. It had 500 to 1,000 residents with an average annual income of $6,500. The average age was between fifty and sixty years, and most residents were retired citrus workers.

The group committed to the PACE-EH thirteen-task process (Figure 13.6), and what they discovered was not, at first glance, a traditional public health issue. Of the major issues identified, the number one issue was a need to decrease violence and unintended injury. The proposed solution? Streetlights so people could walk safely in the evenings. The second issue was a need for sidewalks, and the third was a need for walking paths. It was not until

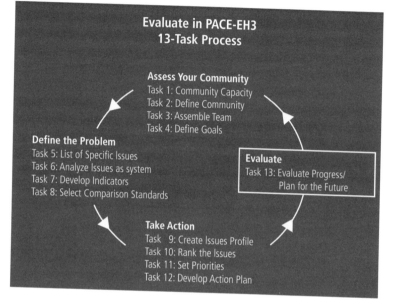

**Evaluate in PACE-EH3
13-Task Process**

Assess Your Community
Task 1: Community Capacity
Task 2: Define Community
Task 3: Assemble Team
Task 4: Define Goals

Define the Problem
Task 5: List of Specific Issues
Task 6: Analyze Issues as system
Task 7: Develop Indicators
Task 8: Select Comparison Standards

Evaluate
Task 13: Evaluate Progress/
Plan for the Future

Take Action
Task 9: Create Issues Profile
Task 10: Rank the Issues
Task 11: Set Priorities
Task 12: Develop Action Plan

Figure 13.6   Evaluation in PACE-EH—13-Task Process.
Source: Centers for Disease Control and Prevention.

much further down the list that traditional public health topics, such as the need for clean water and septic systems, showed up.

So this group, without any money or political power, mobilized and articulated their message, making clear why it's important for Wabasso to be a safe community, and they have raised $1.5 to $2 million in private funds to put in streetlights, sidewalks, walking paths, safer homes, and even septic systems. The *pace-setters*, as they call themselves (because they used the PACE-EH process), walk on their new paths and talk about next steps.[15]

If Wabasso can do this, then people anyplace else can also use this process and figure out the real, underlying risks in their community. They can address these underlying causes, rather than putting Band-Aids on the symptoms, which only make it harder to see the pervasive problem.

When you're just getting started, the town meeting structure, a *charrette*, can bring a community together to galvanize change. The former and current mayors of Elgin, Illinois, Ed Schock and Dave Kaptain, have used this format to deliver information, receive feedback from the community, and generate community buy-in. One key to successful public meetings is to have an organized agenda with opportunities for people to give real and substantial input. Then urge the participants to set their personal biases and interests aside in order to keep a broad view on topics as the discussion proceeds. When Ed Schock was mayor, he set the participants at one such meeting to work with these words:

When I look down the list of people that have volunteered, signed up, and committed to this effort, there is expertise in a variety of areas. We have members of committees; tremendous backgrounds have been brought to the endeavor. As we embark on today's work, my final plea is this: as we apply our individual knowledge and our interests to the task ahead, maintain the broader view. It isn't just about water, clean air, energy use, or reducing energy consumption. It's also about economic development. A community that doesn't have jobs and economic development for its citizens isn't sustainable. What have we achieved? This is just the beginning. We have a long way to go.

Realize

When you are facing a big project, there will be days when the task seems overwhelming and progress seems out of reach. I make lists because I like crossing things off. We all need to see what we've accomplished along the way—the personal triumphs as well as the mission inching toward its eventual goal.

Sometimes the work of change is about keeping a vision alive, keeping it in front of you and talking about it until a critical mass of support moves the idea along. Creating policy is a huge accomplishment, but it is still only a milestone along the way. The issue is not resolved until the policy is implemented and change takes place. However, that does not diminish the accomplishment.

What does it take to keep going in the face of fatigue and disappointment? In our personal lives we can turn to community—to friends, family, and places of worship—to give us strength through the tougher times and to help us celebrate during the joyous moments. In this work you are the leader and it is up to you to create space for people to let out their frustrations when things do not work as planned, and also up to you to welcome and instigate

TAJA SEVELLE ON FEELING LIKE QUITTING

I have had thoughts about quitting Detroit, because sometimes it has been extremely hard. But they are fleeting, and I am not a quitter. There are too many miracles that happen every single day with us, and there are a lot of people depending on this. We have a vision you know, we have a vision of a different world where we aren't looking at unused land. We are looking at land that's used until it's developed. We are looking at walls and at rooftops that have gardens bearing wonderful life-giving fruits and vegetables. We have a vision that our children will think that that's normal to see gardens, and they will be used to seeing them. Then they will ask us what it was like when we had hunger.

celebration along the way. A bit of music and the sharing of a snack, hugs, and kind words can do wonders to raise spirits and rejuvenate people's energy. Keep track of the accomplishments. When people ask, "Have we made any difference?" take out your folder with thank-you notes and e-mails proving that you and your stakeholders have in fact made a difference.

As one of the change effort participants in Elgin, Illinois, said, "We celebrate, celebrate our work; like whatever you do, you got to have a celebration behind it, and so that's what we do."

PROTECT THROUGH PREVENTION

In hospitals, doctors say that patients are either improving or getting worse—there is no such thing as *stable*. I would argue that the same is true of a community. It is either improving or declining. Continual self-evaluation and prioritizing moves the community ever forward.

In order to keep efforts moving, leadership must be replenished. This requires thoughtful and conscious capacity building. As magnetic a leader as you may be, and as fabulous a working team as you may have put together, a time will come when you or someone on your team needs to move on to other areas of focus. Always be grooming future leaders.

Let me close this chapter on action plans with one last example that illustrates the kinds of complexities that our work to design healthy environments faces and the size of the challenge we face. Sacramento is located in California's Central Valley, with the prevailing winds blocked to the east by the Sierra Nevada mountains and with air pollutants trapped over and to the east of the city. The city and its residents have been working hard to clean the air and have made substantial progress in the last few years, but the big issue is how to shorten the distance of people's daily commute. Doing this will require increasing density, which will also increase the numbers of people who are living close to downtown highways, with higher exposure to air pollutants. While the air is likely better in the suburbs, the air quality on the highways is no better during the commute, and commuting carries injury and cardiovascular disease risks. We need to do an all-costs accounting of risks and benefits of these life choices, and the policymakers who are weighing these need to look at the spectrum of risk, not just those within one domain such as air pollution. Health is not just about isolated body parts or discrete environmental media, it is health for our entire body and health for all.

EPILOGUE

NOW IT'S YOUR TURN

Our built environment is the result of design ideas that came before us and were captured in landscape, wood, stone, steel, and concrete. We can use the tools at our disposal to tear down what is no longer useful and create a new vision for ourselves, our children, and our grandchildren. The world can become what we envision when we wield the tools of affectionate expertise as a sculptor wields a mallet and chisel.

WHAT CAN ONE PERSON DO?

One person can effect change by making choices that reflect a group's or a community's core values at each stage where action is possible. We can make healthier personal choices, we can mitigate the causes of disease, and we can reduce waste. We can make healthier personal choices, we can improve the systems that support a healthy lifestyle, and we can mitigate or eliminate the factors that greatly increase our disease risk. Working

at the individual level, the cumulative effect of people making small choices can be measurable. We can choose to take a walk rather than sit at home or to ride a bike rather than drive. We can spend our financial capital to support local businesses or apply market forces when we choose healthier options over junk food. We can choose to move to neighborhoods that are walkable, or we can work with our community to build a neighborhood we want to walk in.

Yang Li, one of Deb Perryman's high school students in Elgin, Illinois, explained her commitment to change this way: "I've got my entire life ahead of me. I don't want to be living in a world where the sky is gray every day and where I have to go downtown and see that brown streak through the sky that's full of pollution. I don't want to have to deal with that for the rest of my life, and I don't want my children or my grandchildren to deal with that. I really want to make a difference now."

Business owners can effect change on as small a scale as deciding what office supplies to purchase, as well as on a larger scale when setting the direction of the strategic plan. Decisions about where the business will be located and what amenities the building should offer can be made with the safety, health, and beauty of the neighborhood in mind. Businesses can choose to have buildings that are safe and attractive, that have accessible bicycle racks, and that provide access to healthy food.

One good decision can lead to another, as a business owner in Oakland, California, found:

> I started getting interested in climate change maybe four or five years ago. It was just a simple little thing in this building that triggered it. We were using all these incandescent bulbs. When we saw how much money the building could save by going to CFL [compact fluorescent lamp] and LED [light-emitting diode] bulbs, we just had to try it. In one month our electricity bill went from $4,500 to $3,100. That was a big savings, so we started thinking, what else can we do? Now we have a green roof, and we put a community garden in the back. There's not a big difference in cost between going with green building materials or green building products and the products that we were used to. So, we're going to be painting all of our hallways with no VOC [volatile organic compound] paint and covering the floors with eco-friendly carpet. We had a contest among ourselves in the building to see how much money we can save. We're making it fun.

Government at its best is the work of people driven to improve the common good, especially over the long term. Although government is often represented as a huge, faceless organization, it has many individuals who want to make their communities better—healthier, more beautiful, and safer. Individuals with access to a podium have the opportunity to educate the public on important

Howard Frumkin Answers the Question, Whose Responsibility Is It?

Our generation and the generation before in the post–World War II era fell in love with the car, with open space, and we had seemingly inexhaustible supplies of gasoline, land, and everything else that it took to build sprawling metropolitan areas. It's time now to change and we know it. We've learned about some of the elements of change that we need, and the younger generation is going to lead us in that direction. My children are very, very tuned into this. Without being taught this, they inherently prefer neighborhoods that are walkable. They like traveling by bicycle, being in the outdoors, and green surroundings in their neighborhoods. My kids are wonderful, but they're probably not unusual in those respects. I think that lots of young people get this issue, and that's a very encouraging sign as we look towards positive change in the future.

issues and tap into people's inherent interest in the betterment of their community.

For example, Elgin's Ed Schock had a very clear idea of what his role as mayor of a changing community needed to be:

> I think what the mayor can do to help accomplish our goals is to be a community voice and advocate—to be constantly bringing the message forward. I was an educator for thirty-four years and I believe in the power of education. I think elected officials have a responsibility to educate the community and answer questions about why this is the direction we should go, why it is in everyone's self-interest. Ultimately people are motivated by their own self-interest—why is this going to be good for me, my family, my neighborhood, my community?

At all levels of our community, we can step in at our comfort level and effect change.

WHO INSPIRES YOU?

We all get ideas from somewhere. What have you been reading? What films do you watch? Whom do you talk with about your hopes for the future? People all over the world are carrying out small, miraculous actions to make our world healthier. Some are highlighted by the media and recognized by awards. Even though recognition is sometimes a good experience, I don't think it's the reason

people actively engage in civic projects, though recognition of accomplishment spreads the word that good work is being done, and encourages others to join in.

An Atlanta high school student, for example, has found inspiration in the work of a recipient of the Nobel Peace Prize:

> Wangari Maathai won the Nobel Peace Prize in 2004 for her work on the Green Belt Movement in Kenya. She is an incredible environmental and political activist and I love how she planted trees in Kenya to prevent soil erosion. I realize that we can do the same and improve our air quality by planting trees right here in Atlanta. We can do that not just in Kenya, not just in Atlanta, but anywhere in the United States if we want to. I like how she says that young people need to become involved in promoting environmental sustainability. We have to instill the idea that protecting the environment is not just our pleasure but is also our duty.

Sometimes little actions are all that are needed even for upstream intervention. I recently met Wendy Slusser, who works at the Venice Family Clinic in Venice, California. "When I look at somebody who is not healthy, I actually don't blame him or her," she said. "I feel that the environment has propelled a large group of our population to be unhealthy. As an example, we had a park that was closed three years ago because of gang activity. I was doing a lot of education for children and families on nutritional choices, but once the park reopened it was really like the one-two punch. People had the nutritional information and a place where they could be active. Suddenly there was a transformation in terms of their weight and overall health."

WHY NOT YOUR COMMUNITY?

The communities I have chosen to write about were not selected because they are perfect examples of healthy built environments. They are highlighted here because they are firmly on the road to becoming healthier and because they have focused on a well-defined problem area in which they have the ability to improve. Moreover, they are only a few of the models across the American landscape; quite a number of other communities are also working on becoming healthier. Each of the communities I discuss has faced barriers and has had to build coalitions to make change. Each has a vision, and none is yet done with realizing that vision. I do not think a community can ever be *done*. So, where else will these ideas take root next? Why not in your community?

WHY NOT NOW?

Sometimes in my field there is no way the public will be satisfied without a serious epidemiological study, but after the time and resources are spent, the study findings may still not have a meaningful impact. A scientific investigation doesn't always lead to an answer. My own experience is that oftentimes large amounts of resources are put into an investigation, and yet in the end we do not know much more than we did at the beginning—which is that community residents across the United States often have inadequate medical and dental care and that the kids are way behind in school, immunizations, and nutrition. Studies take time, time we don't have. I think we are in a Code Blue stage (that is, a true

Taja Sevelle Describes Creating Something Positive in a Shrinking Community

Detroit has over 5,000 acres of unused land in the city proper, so it seems like a great place to demonstrate that this unused land can be a win-win for the people and the community. Where you put in a garden, there is less crime. The owners are "off the street" to till the field. The garden feeds people, the primary goal, but it's also reaching out to the community where people start to talk about things like entrepreneurship, money management, food, and nutrition. So, we created a garden and people can bring their kids. The kids learn how to grow and care for plants. In turn they will be able to teach others.

emergency)—we need to act and not just investigate and pontificate. Often professionals in public health are working outside confirmed science, because by the time the studies are completed and causation is known for sure, it may be too late to act for the best outcome. So, if I had to choose between implementing an epidemiological study and providing care, the study would not be my first choice. People need help, they need it now, and we know the essential actions that can do *good*.

In situations that are less lethal than those I often deal with, one could argue for taking the time to get perfect information before implementation, but Ed Schock disagrees:

> You can't wait till you have perfect knowledge because you'll never have it. Nobody really knows today what all the elements of sustainability are. There is more information and new technologies to develop, so if we think this is going to have an end point where we say, 'Yes, we are a sustainable community,' that's probably not going to happen. What it is going to require ultimately is a change in lifestyle. The way we live our lives on a day-to-day basis is going to determine how sustainable our community is.

It seems I receive a note every few weeks from a former student saying that he or she never goes through town anymore without thinking about how it's laid out. The physical environment tells us what's important—the cars, the bicyclists, or the pedestrians. The more you know about the built environment, the more you think about what that environment is communicating—and you start to listen.

Right now, people shopping for a home are checking for *walkability*, connectedness, and sense of community. The marketplace is starting to turn toward livable communities. I love it when the young parents who have just bought the house down the street are asked by the pediatrician, "Are you in a neighborhood where your child can walk and play with friends?"

You can find out how walkable your community is in relation to your house, or any new house you've been considering, by using a free, online calculator called Walk Score (Figure E.1), at www.walkscore.com. You enter any address, and the service instantly looks up how far you are from the essential services in your community. The system then generates a score from 0 to 100, where 100 is as walkable as an address can be. I am impressed that this idea has gone from something that only those of us engaged in built environment and walkability issues knew about to something every home buyer is aware of. I always smile when I see a Walk Score in a real estate ad, but I have noticed that happens only when it is a good one—no one is bragging about a bad Walk Score.

WHY NOT YOU?

This country's communities need you to change the world. They need you to lead change onward to a world

Figure E.1   A Walk Score is one of the tools for determining the desirability of a location.

Source: Walk Score, Home Page, http://www.walkscore.com. © www.walkscore.com.

where kids are given good healthy food that does not abuse their bodies or our planet. They need you to show that America is not indifferent to the poor and has a higher vision than merely moving wealth to those already wealthy. They need you to show that public health is about creating the conditions where people can become fully realized human beings.

On his wonderful show, Mr. Rogers used to remind children that they are special. In contrast to most current media messages, he told every child that he or she was special not because of what he or she might have, or what goods he or she might consume, or even how he or she might look or how smart he or she might be. Every child is special because every child is unique. Every one of us is special too, because we have the choice before us to accept the gift of service and to do something more than fulfill our own immediate wants and needs.

America needs you and the world needs you. You may already have worked hard for what you have, but you and everyone else will need to work harder. Our communities need you to learn as much about yourself and the world as possible. Science is important, politics is crucial, decisions have lasting impact, but unless they are based on humane ethics and decency, all of them are without value. Remember, without courage and integrity, all other virtues are meaningless. You are about to step into amazing challenges and remarkable opportunities. You have arrived in the nick of time.

NOTES

Preface

1   For more information about designing healthy communities and the public television series, please visit the Media Policy Center's Web sites http://www.designinghealthycommunities.org and http://www.mediapolicycenter.org.

Prologue

1   K. M. Venkat Narayan and others, "Lifetime Risk for Diabetes Mellitus in the United States," *JAMA* (2003), *290*(14), 1884–1890, doi: 10.1001/jama.290.14.1884.

2   S. Keehan and others, "Health Spending Projections Through 2017: The Baby-Boom Generation Is Coming to Medicare," *Health Affairs* (2008), *27*(2), 145–155.

Chapter One

1   C. L. Ogden and others, "Prevalence of High Body Mass Index in US Children and Adolescents, 2007–2008," *JAMA* (2010), *303*(3), 242–249.

2   Centers for Disease Control and Prevention, "County-Specific Diabetes and Obesity Prevalence, 2007" (Nov. 2010), http://www.cdc.gov/obesity/data/trends.html#Race.

3   E. Dunham-Jones and J. Williamson, *Retrofitting Surburbia: Urban Design Solutions for Redesigning Suburbs* (Hoboken, N.J.: Wiley, 2009).

4   S. Rosenbloom, "But Will It Make You Happy?" *New York Times* (Aug. 7, 2010).

5   J. Carroll, "Workers' Average Commute Round-Trip Is 46 Minutes in a Typical Day: Commute to and from Work Is Not Stressful for Most Workers," Gallup News Service (Aug. 24, 2007), http://www.gallup.com/poll/28504/workers-average-commute-roundtrip-minutes-typical-day.aspx.

6   National Institute of Mental Health, "Major Depressive Disorder Among Adults" (July 2010), http://www.nimh.nih.gov/statistics/1MDD_ADULT.shtml.

7   P. J. Carek, S. E. Laibstain, and S. M. Carek, "Exercise for the Treatment of Depression and Anxiety," *International Journal of Psychiatry in Medicine* (2011), *41*(1), 15–28.

Chapter Two

1   Centers for Disease Control and Prevention, "Control of Infectious Diseases, 1900–1999," *Morbidity and Mortality Weekly Report* (July 30, 1999), *48*, 621–629.

2   W. C. Knowler and others, "Reduction in the Incidence of Type 2 Diabetes with Lifestyle Intervention or Metformin," *New England Journal of Medicine* (2002), *346*(6), 393–403.

3   W. King and others, "The Relationship Between Convenience of Destinations and Walking Levels in Older Women," *American Journal of Health Promotion* (2003), *18*(1), 74–82.

4   "Environmental Factors the Major Cause of Cancer," *Environmental News Service* (June 28, 2004), http://www.ens-newswire.com/ens/jun2004/2004-06-28-02.asp.

5   E. S. Morgan (ed.), *Not Your Usual Founding Father: Selected Readings from Benjamin Franklin* (New Haven, Conn.: Yale University Press, 2006).

6   B. P. Lanphear and others, "Low-Level Environmental Lead Exposure and Children's Intellectual Function: An International Pooled Analysis," *Environmental Health Perspectives* (2005), *113*(7), 894–899; J. Schwartz, "Low-Level Lead Exposure and Children's IQ: A Meta-Analysis and Search for a Threshold," *Environmental Research* (1994), *65*(1), 42–55.

7   S. D. Grosse, T. D. Matte, J. Schwartz, and R. J. Jackson, "Economic Gains Resulting from the Reduction in Children's Exposure to Lead in the United States," *Environmental Health Perspectives* (2002), *110*(6), 563–569.

8   Centers for Disease Control and Prevention, *Fourth National Report on Human Exposure to Environmental Chemicals* (Atlanta: U.S. Department of Health and Human Services, Centers for Disease Control and Prevention, 2009), p. 212, http://www.cdc.gov/exposure report/pdf/FourthReport.pdf.

9   U.S. Environmental Protection Agency, Office of Air and Radiation, *Report to Congress on Indoor Air Quality*, Vol. 2. *Assessment and Control of Indoor Air Pollution*, EPA 400-1-89-001C (Washington, D.C.: U.S. Environmental Protection Agency, 1989), pp. I, 4–14.

10 M. Orenstein and others, "Safe Routes to School Safety and Mobility Analysis," Paper UCB-TSC-RR-2007-1 (Berkeley, Calif.: UC Berkeley Traffic Safety Center, Apr. 1, 2007); R. Ewing, W. Schroeer, and W. Greene, "School Location and Student Travel Analysis of Factors Affecting Mode Choice," *Transportation Research Record* (2004), no. 1895, 55–63.

11 E. A. Finkelstein, J. G. Trogdon, J. W. Cohen, and W. Dietz, "Annual Medical Spending Attributable to Obesity: Payer- and Service-Specific Estimates," *Health Affairs* (July 29, 2009), http://content.healthaffairs.org/cgi/content/abstract/hlthaff.28.5.w822v2.

12 R. Hammond and R. Levine, "The Economic Impact of Obesity in the United States."

Diabetes, Metabolic Syndrome and Obesity: Targets and Therapy (Aug. 17, 2010), *3*, 285–295.

13 J. Block, R. A. Scribner, and K. B. DeSalvo, "Fast Food, Race/Ethnicity, and Income: A Geographic Analysis," *American Journal of Preventive Medicine* (2004), *27*(3), 211–217; this article describes a 2001 study of fast-food restaurants in New Orleans.

14 H. Frumkin, "Urban Sprawl and Public Health," *Public Health Reports* (May–June 2002), *117*, 201–217, http://www.cdc.gov/healthyplaces/articles/Urban%20Sprawl%20and%20Public%20Health%20-%20PHR.pdf.

15 J. L. Atkinson and others, "The Association of Neighborhood Design and Recreational Environments with Physical Activity," *American Journal of Health Promotion* (2005), *19*(4), 304–309; B. Giles-Corti and others, "Increasing Walking: How Important Is Distance to, Attractiveness, and Size of Public Open Space," *American Journal of Preventive Medicine* (2005), *28*(Suppl. 2), 169–176; L. D. Frank and G. Pivo, "Impacts of Mixed Use and Density on Utilization of Three Modes of Travel: Single-Occupant Vehicle, Transit, and Walking," *Transportation Research Record* (Apr. 1995), no. 1466, 44–52; D. A. Cohen and others, "Contribution of Public Parks to Physical Activity," *American Journal of Public Health* (2007), *97*, 509–514; J. N. Roemmich and others, "Association of Access to Parks and Recreational Facilities with the Physical Activity of Young Children." *Preventive Medicine* (2006), *43*, 437–441.

16 T. Takano, K. Nakamura, and M. Watanabe, "Urban Residential Environments and Senior Citizens' Longevity in Megacity Areas: The Importance of Walkable Green Spaces," *Journal of Epidemiology & Community Health* (Dec. 2002), *56*, 913–918.

17 Centers for Disease Control and Prevention, *CDC Plague Home Page* (2009), http://www.cdc.gov/ncidod/dvbid/plague/.

18 Tobacco-Free California, "Consumption of Tobacco Products Declines Dramatically" (2008), http://www.tobaccofreeca.com/ca_success.html.

19 "Pollution from Asia Circles Globe at Stratospheric Heights," *ScienceDaily* (Mar. 26, 2010), http://www.sciencedaily.com/releases/2010/03/100325143047.htm.

20 U.S. Environmental Protection Agency, Office of Transportation and Air Quality, "Solutions That Reduce Pollution" (July 9, 2007), http://www.epa.gov/otaq/invntory/overview/solutions/vech_engines.htm.

21 U.S. Environmental Protection Agency, Office of Transportation and Air Quality, *EPA and NHTSA Finalize Historic National Program to Reduce Greenhouse Gases and Improve Fuel Economy for Cars and Trucks* (Apr. 2010), http://www.epa.gov/oms/climate/regulations/420f10014.pdf.

22 World Health Organization, "Constitution of the World Health Organization" (1948/2006), http://www.who.int/governance/eb/who_constitution_en.pdf.

23 L. Szabo, "Number of Americans Taking Antidepressants Doubles," *USA Today* (Aug. 4,

2009), http://www.usatoday.com/news/health/2009-08-03-antidepressants_N.htm.

24 I. B. Kawachi, I., B. Kennedy, K. Lochner, and D. Prothrow-Stith, "Social Capital, Income Inequality, and Mortality," *American Journal of Public Health* (1997), *87*(9), 1491–1498.

25 N. Mustafa, "What About Gross National Happiness?" *Time* (Jan. 10, 2005), http://www.time.com/time/health/article/0,8599,1016266,00.html.

26 J. Lichfield, "Sarkozy's Happiness Index Is Worth Taking Seriously," *The Independent* (Sept. 20, 2009), http://www.independent.co.uk/opinion/commentators/john-lichfield/john-lichfield-sarkozys-happiness-index-is-worth-taking-seriously-1790323.html.

27 J. H. Kunstler, "Home from Nowhere," *Atlantic Monthly* (Sept. 1996), p. 1.

Chapter Three

1 See, for example, Shay Salomon's Web page at http://www.resourcesforlife.com/library/people/shay-salomon/.

2 See, for example, R. Florida, *The Rise of the Creative Class* (New York: Basic Books, 2002).

3 R. D. Putnam, *Bowling Alone: The Collapse and Revival of American Community* (New York: Simon & Schuster, 2000).

Chapter Four

1 J. Neff, "Is Digital Revolution Driving Decline in U.S. Car Culture? Shift Toward Fewer Young Drivers Could Have Repercussions for All Marketers," *Advertising Age* (May 31, 2010), http://adage.com/print?article_id=144155.

2 F. Kent and K. Madden, "The Origin of the Power of Ten," Project for Public Spaces, http://www.pps.org/articles/poweroften/.

3 United Health Foundation, "Colorado (2009)," *America's Health Rankings*, http://www.americashealthrankings.org/yearcompare/2008/2009/CO.aspx.

Chapter Five

1 In addition to the sources listed in the remaining notes for this chapter, the following were used in writing this chapter: S. B. Buntin, "Prairie Crossing," Unsprawl Case Study (n.d.), http://www.terrain.org/unsprawl/9/; C. Ellis, "The New Urbanism: Critiques and Rebuttals," *Journal of Urban Design* (2002), *7*(3), 261–291; V. Ranney, K. Kirley, and M. Sands, *Building Communities with Farms: Insights from Developers, Architects and Farmers on Integrating Agriculture and Development* (Grayslake, Ill.: Liberty Prairie Foundation, 2010), http://www.prairiecrossing.com/libertyprairiefoundation/cwf.pdf; National Geographic, *The New Suburb? Comparing New-Urbanist and Sprawl Suburbs* (n.d.), http://www.nationalgeographic.com/features/00/earthpulse/sprawl/gallery1.html.

2 A. Duany, E. Plater-Zyberk, and J. Speck, *Suburban Nation: The Rise of Sprawl and the Decline of the American Dream* (New York: North Point Press, 2000).

3 Prairie Crossing, Prairie Crossing Web site, accessed on May 2010, http://www.prairiecrossing.com/pc/site/about-us.html.

4 Adapted from Prairie Crossing, Prairie Crossing Web site, accessed on May 2010, http://www.prairiecrossing.com/pc/site/guiding-principles.html.

5 F. B. Hu and others, "Adiposity as Compared with Physical Activity in Predicting Mortality Among Women," *New England Journal of Medicine* (2004), *351*(26), 2694–2703, http://www.nejm.org/doi/full/10.1056/NEJMoa042135.

6 M. Thao, *Safe Routes to School: Seward Montessori School: Encouraging Walking and Biking to School* (St. Paul, Minn.: Wilder Research, Dec. 2010), http://www.wilder.org/download.0.html?report=2384.

7 K. R. Fox, "The Influence of Physical Activity on Mental Well-Being," *Public Health Nutrition* (1999), *2*(3A), 411–418; S. N. Young, "How to Increase Serotonin in the Human Brain Without Drugs," *Journal of Psychiatry & Neuroscience* (2007), *32*(6), 394–399; C. Ernst and others, "Antidepressant Effects of Exercise: Evidence for an Adult-Neurogenesis Hypothesis?" *Journal of Psychiatry & Neuroscience* (2006), *31*(2), 84–92.

8 Centers for Disease Control and Prevention, "Barriers to Children Walking to or from School—United States, 2004," *Morbidity and Mortality Weekly Report* (Sept. 30, 2005), *54*, 949–952.

9 Pacific Gas and Electric Company, *Daylighting in Schools: An Investigation into the Relationship Between Daylighting and Human Performance*

(1999), http://www.coe.uga.edu/sdpl/research/daylightingstudy.pdf.

10 U.S. Department of Agriculture, *FY 2006 Budget Summary and Annual Performance Plan* (Washington, D.C.: U.S. Department of Agriculture, n.d.).

Chapter Six

1 D. Nozzi, "Charleston Mayor Joseph P. Riley," *Walkable Streets* (Nozzi's Web site), (Feb. 23, 1997), http://www.walkablestreets.com/riley.htm; a compilation of excerpts from a speech made by Joe Riley.

2 Nozzi, "Charleston Mayor Joseph P. Riley."

3 Nozzi, "Charleston Mayor Joseph P. Riley."

4 City of Charleston, "Mayor Joseph P. Riley, Jr—Biography" (2011), http://charlestoncity.info/dept/content.aspx?nid=495.

5 P. Bowers, "Stats and Facts," *My Charleston* (Dec. 2010), http://www.mycharlestononline.com/news/2010/aug/31/stats-facts/.

6 City of Charleston, "Mayor Joseph P. Riley, Jr—Biography."

7 D. Florio, "Charleston's Mayor—Joe Riley," *Southern Living Magazine* (Mar. 2010).

Chapter Seven

1 Centers for Disease Control and Prevention, "Illinois: Burden of Chronic Diseases" (Atlanta: Centers for Disease Control and Prevention,

2009); D. Blakley, "Elgin Declared Fattest City in Illinois: But Distinction Sparks Little Interest in 'Fitness Grants,'" CBS Chicago (Aug. 13, 2009), http://cbs2chicago.com/local/elgin.fattest.city.2.1127034.html.

2 Kane County Health Department, *Vital Signs: 2009 Report to the Community*, Annual Report (n.d.), http://www.kanehealth.com/PDFs/AnnualReports/KCHDAnnualReport09.pdf.

3 Carpet America Recovery Effort, *2007 Annual Report* (n.d.), http://www.carpetrecovery.org/pdf/annual_report/07_CARE-annual-rpt.pdf.

Chapter Eight

1 Colorado Children's Campaign, *Childhood Obesity in Colorado: A Growing Problem* (Denver: Colorado Children's Campaign, July 2007).

2 U.S. Department of Health and Human Services, Assistant Secretary for Planning and Evaluation, *Childhood Obesity* (n.d.), http://aspe.hhs.gov/health/reports/child_obesity/.

3 Colorado Department of Public Health and Environment, *Colorado Child Health Survey*, (Denver: Colorado Department of Public Health and Environment, 2007).

4 Centers for Disease Control and Prevention, *Obesity and Overweight: Colorado* (Mar. 3, 2011), http://www.cdc.gov/obesity/stateprograms/fundedstates/colorado.html.

5 Centers for Disease Control and Prevention, *Obesity and Overweight: Colorado*.

6 G. H. Berken, D. O. Weinstein, and W. C. Stern, "Weight Gain: A Side-Effect of Tricyclic Antidepressants," *Journal of Affective Disorders* (1984) 7(2), 133–138.

7 Centers for Disease Control and Prevention, *Childhood Obesity and Overweight: Data and Statistics* (2011), http://www.cdc.gov/obesity/childhood/data.html.

8 U.S. Department of Health and Human Services, Office of the Surgeon General, *The Surgeon General's Call to Action to Prevent and Decrease Overweight and Obesity* (Washington, D.C.: U.S. Government Printing Office, 2001), http://www.surgeongeneral.gov/topics/obesity/; also see Centers for Disease Control and Prevention, *Obesity and Overweight: Causes and Consequences* (Dec. 7, 2009), http://www.cdc.gov/obesity/causes/index.html, and *Childhood Overweight and Obesity* (Mar. 3, 2011), http://www.cdc.gov/obesity/childhood/index.html.

9 C. Leinberger, *The Option of Urbanism: Investing in a New American Dream* (Washington, D.C.: Island Press, 2009).

10 H. Twaddle, F. Hall, and B. Bracic, "Latent Bicycle Commuting Demand and Effects of Gender on Commuter Cycling and Accident Rates," *Transportation Research Record* (2010), no. 2190, 28–36, http://trb.metapress.com/content/j85x8g66k1545860/?p=ee71b629883d413cba017bda6bf34468&pi=3&referencesMode=Show.

11 American Association of State Highway and Transportation Officials, "President Signs SAFETEA-LU: Safe, Accountable, Flexible,

Efficient Transportation Equity Act for the 21st Century—A Legacy for Users," AASHTO Reauthorization Update (June 2010), http://www.transportation1.org/aashtonew/.

Chapter Nine

1 Port of Oakland, "The Port and You" (June 2010), http://www.portofoakland.com/portnyou/history.asp.

2 California Environmental Protection Agency, Air Resources Board, "Appendix A: Emission Inventory Summary," in *Diesel Particulate Matter Health Risk Assessment for the West Oakland Community*, http://www.arb.ca.gov/ch/communities/ra/westoakland/documents/appendixa_final.pdf.

3 California Environmental Protection Agency, Air Resources Board, "Appendix A: Emission Inventory Summary."

4 L. Feldstein, *General Plans and Zoning: A Toolkit for Building Healthy, Vibrant Communities*, a report produced by Planning for Healthy Places at Public Health Law & Policy in partnership with the California Department of Health Services through the California Nutrition Network for Healthy, Active Families, http://www.phlpnet.org/healthy-planning/products/general-plans-and-zoning.

5 Alameda County Public Health Department, *Death from All Causes* (2003), http://www.acphd.org/axbycz/admin/datareports/4achsr03_death_all&leading.pdf; Alameda County Public Health Department, *Life and Death from Unnatural Causes: Health and Social Inequity in*

Alameda County (Apr. 18, 2008), http://www.acphd.org/healthequity/reports/index.htm.

6 M. Palaniappan, D. Wu, and J. Kohleriter, *Clearing the Air: Reducing Diesel Pollution in West Oakland* (Oakland, Calif.: Pacific Institute and the Coalition for West Oakland Revitalization, Nov. 2003).

7 M. Palaniappan and S. Prakash, *Paying with Our Health: The Real Cost of Freight Transport in California* (Oakland, Calif.: Pacific Institute, Nov. 2006).

8 J. M. MacDonald and others, "The Effect of Light Rail Transit on Body Mass Index and Physical Activity," *American Journal of Preventive Medicine* (2010), *39*(2), 105–112.

9 Alameda County Public Health Department, *Life and Death from Unnatural Causes*.

10 R. McConnell and others, "Asthma in Exercising Children Exposed to Ozone: A Cohort Study," *Lancet* (2002), *359*(9304), 386–391.

11 S. Costa and others, *Neighborhood Knowledge for Change: The West Oakland Environmental Indicators Project* (Oakland, Calif.: Pacific Institute, 2002).

Chapter Ten

1 U.S. Census Bureau, *State & County QuickFacts* (Aug. 2010), http://quickfacts.census.gov/qfd/states/26/2622000.html.

2 R. French and M. Wilkinson, "Leaving Michigan Behind: Eight-Year Population Exodus Staggers State," *Detroit News*, Apr. 2, 2009.

3 J. A. Lozano, "Detroit Eclipses Houston as Fattest City," *AP Online* (Jan. 3, 2004), http://www.highbeam.com/doc/1P1-89027914.html.

4 "2009 Fittest/Fattest Cities: Our 11th Annual Survey Revealed Some Surprises," *Men's Fitness* (Dec. 2010), http://www.mensfitness.com/lifestyle/215.

5 A. Bishaw and J. Semega, U.S. Census Bureau, *Income, Earnings, and Poverty: Data from the 2007 American Community Survey* (Washington, D.C.: U.S. Government Printing Office, 2008), http://www.census.gov/prod/2008pubs/acs-09.pdf.

6 "Detroit's Unemployment Rate Is Nearly 50%, According to the Detroit News." *Huffington Post* (Dec. 16, 2009), http://www.huffingtonpost.com/2009/12/16/detroits-unemployment-rat_n_394559.html?view=print.

7 M. Luo, "At Closing Plant, Ordeal Included Heart Attacks," *New York Times* (Feb. 24, 2010), http://www.nytimes.com/2010/02/25/us/25stress.html.

8 K. B. Kavanaugh, "First—of 30—Green, Modular Homes Going Up at Alter Commons This Week," *Detroit Development News* (May 12, 2009), http://www.modeldmedia.com/devnews/altercommons19109.aspx.

9 Eastern Market Corporation, "Top Ten Reasons Why Eastern Market Is and Will Remain Detroit's Most 'True Green' Neighborhood" (2007), http://www.detroiteasternmarket.com/page.php?p=4&s=90.

10 Detroit 300 Conservancy, "Campus Martius Park" (2006), http://www.campusmartiuspark.org/history.htm.

Chapter Eleven

1   J. Blevins, "Bicyclists Want to Derail Black Hawk's Ban," *Denver Post* (June 15, 2010), http://www.denverpost.com/ci_15298056.

Chapter Twelve

1   M. Josephson, "Using All Your Strength," http://charactercounts.org/michael/2007/10/using_all_your_strength.html; this story is adapted from D. Wolpe, *Teaching Your Children About God* (New York: Harper Perennial, 1995), p 214.

Chapter Thirteen

1   Architecture 2030, "The 2030 Challenge" (2010), http://www.architecture2030.org/2030_challenge/the_2030_challenge.

2   Port of Los Angeles, "About the Port" (Dec. 2010), http://www.portoflosangeles.org/about/facts.asp.

3   National Highway Traffic Safety Administration, *Fatality Analysis Reporting System Encyclopedia* (2009), http://www-fars.nhtsa.dot.gov/Vehicles/VehiclesAllVehicles.aspx.

4   Congress for the New Urbanism, "San Francisco's Octavia" (n.d.), http://www.cnu.org/highways/sfoctavia.

5   Centers for Disease Control and Prevention, *Then and Now—Barriers and Solutions* (2005), http://www.cdc.gov/nccdphp/dnpa/kidswalk/then_and_now.htm; Centers for Disease Control and Prevention, "Physical Activity and the Health of Young People," Fact Sheet (Atlanta: Centers for Disease Control and Prevention, 2004); U.S. Department of Transportation, Federal Highway Administration, *Safe Routes to School* (Dec. 2010), http://safety.fhwa.dot.gov/saferoutes/.

6   American Society of Landscape Architects, *Livable Communities* (June 2010), http://www.asla.org/ContentDetail.aspx?id=23268.

7   R. Florida, *The Rise of the Creative Class* (New York: Basic Books, 2002).

8   Smart Growth America, *Social Equity* (June 2010), http://www.smartgrowthamerica.org/socialequity.html.

9   D. Nozzi, "Charleston Mayor Joseph P. Riley," *Walkable Streets* (Nozzi's Web site) (Feb. 23, 1997), http://www.walkablestreets.com/riley.htm.

10 M. L. Bell and D. L. Davis, "Reassessment of the Lethal London Fog of 1952: Novel Indicators of Acute and Chronic Consequences of Acute Exposure to Air Pollution," *Environmental Health Perspectives* (2001), *109* (Suppl. 3), 389–394, http://ehpnet1.niehs.nih.gov/docs/2001/suppl-3/389-394bell/abstract.html.

11 California Environmental Protection Agency, Air Resources Board, *Key Events in the History of Air Quality in California* (July 2010), http://www.arb.ca.gov/html/brochure/history.htm.

12 Institute for Energy Research, "China Set Records in 2009" (Jan. 21, 2010), http://www.instituteforenergyresearch.org/2010/01/21/china-set-records-in-2009-what%E2%80%99s-in-store-for-2010/.

13 D. Bates, "Great Crawl of China," *Mail Online* (Aug. 25, 2010), http://www.dailymail.co.uk/news/worldnews/article-1306058/China-traffic-jam-enters-11th-day-officials-admit-weeks.html.

14 "Off the Rails?" *The Economist* (Mar. 29, 2011), http://www.economist.com/node/18488554.

15 H. Falk, *Healthy Places*, PowerPoint presentation (Mar. 2009), http://www.cphfoundation.org/documents/DrHenryFalk.EnviroHealth_000.pdf.

INDEX

Illustrations are indicated by *fig*.
Photographs in the full-color
Portfolio section are indicated by
plate number (Plate 15).